FUNDAMENTALS OF HYDROLOGY

Water is essential to sustain life on earth and an understanding of water quantity and quality issues is vital for anyone involved in environmental management. *Fundamentals of Hydrology* provides an absorbing and comprehensive introduction to how fresh water moves on and around the planet and how humans affect the quantity and quality of water available to them.

The book consists of three parts, each of fundamental importance in the understanding of hydrology:

- The first section deals with processes within the hydrological cycle, our understanding of them, and how to measure and estimate the amount of water within each process.
- The second section is concerned with the assessment of important hydrological parameters such as streamflow and water quality. It describes techniques used by practising hydrologists in the assessment of water resources.
- The final section of the book draws together the first two parts to discuss water quality and issues of change that impact upon hydrology.

Fundamentals of Hydrology is a lively and accessible introduction to the study of hydrology at university level and will give undergraduates an understanding of hydrological processes, knowledge of the techniques used to assess water resources and an up-to-date overview of hydrology in a changing world. Throughout the text examples and case studies from around the world are used to explain ideas and techniques clearly. Short chapter summaries, essay questions, guides to further reading and a glossary are also included.

Tim Davie currently works as a hydrologist with Landcare Research in New Zealand. Previously he lectured in Environmental Science and Geography at Queen Mary, University of London, for nine years until 2001.

ROUTLEDGE FUNDAMENTALS OF PHYSICAL GEOGRAPHY SERIES

Series Editor: John Gerrard

This new series of focused, introductory text books presents comprehensive, up-to-date introductions to the fundamental concepts, natural processes and human/environmental impacts within each of the core physical geography sub-disciplines. Uniformly designed, each volume contains student-friendly features: plentiful illustrations, boxed case studies, key concepts and summaries, further reading guides and a glossary.

Already published:
Fundamentals of Biogeography
Richard John Huggett

Fundamentals of Soils
John Gerrard

Fundamentals of Hydrology
Tim Davie

Forthcoming:
Fundamentals of Geomorphology
Richard John Huggett
(December 2002)

FUNDAMENTALS OF HYDROLOGY

Tim Davie

Routledge Fundamentals of Physical Geography

London and New York

First published 2003
by Routledge
11 New Fetter Lane, London EC4P 4EE

Simultaneously published in the USA and Canada
by Routledge
29 West 35th Street, New York, NY 10001

Routledge is an imprint of the Taylor & Francis Group

Typeset in Garamond by Keystroke, Jacaranda Lodge, Wolverhampton
Printed and bound in Great Britain by St Edmundsbury Press, Bury St Edmunds, Suffolk

British Library Cataloguing in Publication Data
A catalogue record for this book is available from the British Library

Library of Congress Cataloging in Publication Data
A catalog record for this book has been requested

ISBN 0–415–22028–9 (hbk)
ISBN 0–415–22029–7 (pbk)

In memory of Dean Stewart (1963–2000)
Man of God, fisherman, soil scientist (latent hydrologist) and a true friend,
and for Chris, Katherine and Sarah

CONTENTS

SERIES EDITOR'S PREFACE

We are presently living in a time of unparalleled change and when concern for the environment has never been greater. Global warming and climate change, possible rising sea levels, deforestation, desertification, and widespread soil erosion are just some of the issues of current concern. Although it is the role of human activity in such issues that is of most concern, this activity affects the operation of the natural processes that occur within the physical environment. Most of these processes and their effects are taught and researched within the academic discipline of physical geography. A knowledge and understanding of physical geography, and all it entails, is vitally important.

It is the aim of this *Fundamentals of Physical Geography Series* to provide, in five volumes, the fundamental nature of the physical processes that act on or just above the surface of the earth. The volumes in the series are *Climatology*, *Geomorphology*, *Biogeography*, *Hydrology* and *Soils*. The topics are treated in sufficient breadth and depth to provide the coverage expected in a *Fundamentals* series. Each volume leads into the topic by outlining the approach adopted. This is important because there may be several ways of approaching individual topics. Although each volume is complete in itself, there are many explicit and implicit references to the topics covered in the other volumes. Thus, the five volumes together provide a comprehensive insight into the totality that is Physical Geography.

The flexibility provided by separate volumes has been designed to meet the demand created by the variety of courses currently operating in higher education institutions. The advent of modular courses has meant that physical geography is now rarely taught, in its entirety, in an 'all-embracing' course but is generally split into its main components. This is also the case with many Advanced Level syllabuses. Thus students and teachers are being frustrated increasingly by lack of suitable books and are having to recommend texts of which only a small part might be relevant to their needs. Such texts also tend to lack the detail required. It is the aim of this series to provide individual volumes of sufficient breadth and depth to fulfil new demands. The volumes should also be of use to sixth form teachers where modular syllabuses are also becoming common.

Each volume has been written by higher education teachers with a wealth of experience in all aspects of the topics they cover and a proven ability in presenting information in a lively and interesting way. Each volume provides a comprehensive coverage of the subject matter using clear text divided into easily accessible sections and subsections. Tables, figures and photographs are used where appropriate as well as boxed case studies and summary notes. References to important previous studies and results are included but are used sparingly to avoid overloading the text. Suggestions for further reading are also provided. The main target readership is introductory level undergraduate students of physical geography or environmental science, but there will be much of interest to students from other disciplines and it is also hoped that sixth form teachers will be able to use the information that is provided in each volume.

John Gerrard

AUTHOR'S PREFACE

It is the presence or absence of water that by and large determines how and where humans are able to live. This in itself makes water an important compound, but when you add in that the availability of water varies enormously in time and space, and that water is an odd substance in terms of its physical and chemical properties, it is possible to see that water is a truly extraordinary substance worthy of study at great length. To study **hydrology** is to try and understand the distribution and movement of fresh water around the globe. It is of fundamental importance to a rapidly growing world population that we understand the controls on availability of fresh water. To achieve this we need to know the fundamentals of hydrology as a science. From this position it is possible to move forward towards the management of water resources to benefit people in the many areas of the world where water availability is stressed.

There have been, and are, many excellent textbooks on hydrology. This book does not set out to eclipse all others, rather it is an attempt to fit into a niche that the author has found hard to fill in his teaching of hydrology in an undergraduate Physical Geography and Environmental Science setting. It aims to provide a solid foundation in the fundamental concepts needed to be understood by anybody taking the study of hydrology further. These fundamental concepts are: an understanding of process; an understanding of measurement and estimation techniques; how to interpret and analyse hydrological data; and some of the major issues of change confronting hydrology. One particular aspect that the author has found difficult to find within a single text has been the integration of water quantity and quality assessment; this is attempted here. The book is aimed at first- and second-year undergraduate students.

This book also aims to provide an up-to-date view on the fundamentals of hydrology, as instrumentation and analysis tools are changing rapidly with advancing technology. As an undergraduate, studying physical geography during the 1980s, an older student once remarked to me on the wisdom of studying hydrology. There will be very little need for hydrologists soon, was his line of thought, as computers will be doing all the hydrological analysis necessary. In the intervening twenty years there has been a huge growth in the use of computers, but fortunately his prediction has turned out to be incorrect. There is a great need for hydrologists – to interpret the mass of computer generated information, if nothing else. Hydrology has always been a fairly numerate discipline and this has not changed, but it is important that hydrologists understand the significance of the numbers and the fundamental processes underlying their generation.

There is an undoubted bias in this book towards the description of hydrology in humid, temperate regions. This is a reflection of two factors: the author's main research being in the UK and New Zealand, and that the majority of hydrological research has been carried out in humid and temperate environments. Neither of these is an adequate excuse to ignore arid regions or those dominated by snow and ice melt, and I have tried to incorporate some description of processes relevant to these environs. The book is an attempt to look at the fundamentals of hydrology irrespective of region or physical environment, but it is inevitable that some bias does creep in; I hope it is not to the detriment of the book overall.

There are many people whom I would like to thank for their input into this book. In common with many New Zealand hydrologists it was Dave Murray who sparked my initial interest in the subject and has provided many interesting discussions since. At the University of Bristol, Malcolm Anderson introduced and guided me in the application of modelling as an investigative technique. Since then numerous colleagues and hydrological acquaintances have contributed enormously in enhancing my understanding of hydrology. I thank them all. Keith Smith initially suggested I write this text, I think that I should thank him for that! The reviewers of my very rough draught provided some extremely constructive and useful criticism, which I have tried to take on board in the final version. I would particularly like to thank Dr Andrew Black from the University of Dundee who commented on the initial proposal and suggested the inclusion of the final chapter. Thanks to Ed Oliver who drew many of the diagrams. My wife Chris, and daughters Katherine and Sarah, deserve fulsome praise for putting up with me as I worried and fretted my way past many a deadline while writing this.

Tim Davie
London, December 2001

1

HYDROLOGY AS A SCIENCE

Quite literally hydrology is 'the science or study of' ('logy' from Latin *logia*) 'water' ('hydro' from Greek *hudor*). However, contemporary hydrology does not study all the properties of water. Modern hydrology is concerned with the distribution of water on the surface of the earth and its movement over and beneath the surface, and through the atmosphere. This wide-ranging definition suggests that all water comes under the remit of a hydrologist, while in reality it is the study of fresh water that is of primary concern. The study of the saline water on earth is carried out in oceanography.

When studying the distribution and movement of water it is inevitable that the role of human interaction comes into play. Although human needs for water are not the only motivating force in a desire to understand hydrology they are probably the largest. This book attempts to integrate the physical processes of hydrology with an understanding of human interaction with fresh water. The human interaction can take the form of water quantity problems (e.g. over extraction of **groundwater**) or quality issues (e.g. disposal of pollutants).

> Water is among the most essential requisites that nature provides to sustain life for plants, animals and humans. The total quantity of fresh water on earth could satisfy all the needs of the human population if it were evenly distributed and accessible.
>
> (Stumm, 1986: 201)

Although written nearly twenty years ago, the views expressed by Stumm are still apt today. The real point of Stumm's statement is that water on earth is not evenly distributed and is not evenly accessible. It is the purpose of hydrology as a pure science to explore those disparities and try and explain them. It is the aim of hydrology as an applied science to take the knowledge of why any disparities exist and try to lessen the impact of them. There is much more to hydrology than just supplying water for human needs (e.g. studying floods as natural hazards; the investigation of lakes and rivers for ecological habitats), but analysis of this quotation gives good grounds for looking at different approaches to the study of hydrology.

The two main pathways to the study of hydrology come from engineering and geography, particularly the earth science side of geography. The earth science approach comes from the study of landforms (**geomorphology**) and is rooted in a history of explaining the processes that lead to water moving around the earth and to try to understand spatial links between the processes. The engineering approach tends to be a little more practically based and is looking towards finding solutions to problems posed by water moving (or not moving) around the earth. In reality there are huge areas of overlap between the two and it is often difficult to separate them, particularly when you enter into hydrological research. At an undergraduate level, however, the difference manifests itself through earth science hydrology being more descriptive and engineering hydrology being more numerate.

The approach taken in this book is more towards the earth science side, a reflection of the author's training and interests, but it is inevitable that there is considerable crossover. There are parts of the book that describe numerical techniques of fundamental importance to any practising hydrologist from whatever background, and it is hoped that the book can be used by all undergraduate students of hydrology.

Throughout the book there are highlighted case studies to illustrate different points made in the text. The case studies are drawn from research projects or different hydrological events around the world and are aimed at reinforcing the text elsewhere in the same chapter. Where appropriate there are highlighted worked examples illustrating the use of a particular technique on a real data set.

IMPORTANCE OF WATER

Water is the most common substance on the surface of the earth, with the oceans covering over 70 per cent of the planet. Water is one of the few substances that can be found in all three states (i.e. gas, liquid and solid) within the earth's climatic range. The very presence of that water in the oceans and atmosphere makes it possible for the earth to have a climate that is habitable for life forms: water acts as a *climate ameliorator*. The movement of water between gas, liquid and solid phases is vital for the transfer of energy around the globe: moving energy from the equatorial regions towards the poles. The low viscosity of water makes it an extremely efficient transport agent, whether through international shipping or river and canal navigation. These characteristics can be described as *physical properties* of water.

The *chemical properties* of water are equally important for our everyday existence. Water is one of the

best solvents naturally occurring on the planet. This makes water vital for cleanliness: we use it for washing but also for the disposal of pollutants. The solvent properties of water allow the uptake of vital nutrients from the soil and into plants; this then allows the transfer of the nutrients within a plant's structure. The ability of water to dissolve gases such as oxygen allows life to be sustained within bodies of water such as rivers, lakes and oceans.

The capability of water to support life goes beyond bodies of water; the human body is composed of around 60 per cent water. The majority of this water is within cells, but there is a significant proportion (around 34 per cent) that moves around the body carrying dissolved chemicals which are vital for sustaining our lives (Ross and Wilson, 1981). Our bodies can store up energy reserves that allow us to survive without food for weeks but not more than days without water.

There are many other ways that water affects our very being. In places such as Norway, parts of the USA and New Zealand energy generation for domestic and industrial consumption is through hydro-electric schemes, harnessing the combination of water and gravity in a (by and large) sustainable manner. Water plays a large part in the spiritual lives of millions of people. In Christianity baptism with water is a powerful symbol of cleansing and God offers 'streams of living water' to those who believe (John 7:38). In Islam there is washing with water before entering a mosque for prayer. In Hinduism bathing in the sacred Ganges provides a religious cleansing. Many other religions have water playing an important part in sacred texts and rituals.

Water is important because it underpins our very existence: it is part of our physical, material and spiritual lives. The study of water would therefore also seem to underpin our very existence. Before expanding further on the study of hydrology it is first necessary to step back and take a closer look at the properties of water briefly outlined above. Even though water is the most common substance found on the earth's surface it is also one of the strangest. Many of these strange properties help to contribute to its importance in sustaining life on earth.

Physical and chemical properties of water

A water molecule consists of two hydrogen atoms bonded to a single oxygen atom (Figure 1.1). The connection between the atoms is through **covalent bonding**: the sharing of an electron from each atom to give a stable pair. This is the strongest type of bonding within molecules and is the reason why water is such a robust compound (i.e. it does not break down into hydrogen and oxygen easily). The robustness of the water molecule means that it stays as a water molecule within our atmosphere because there is not enough energy available to break the covalent bonds and create separate oxygen and hydrogen molecules.

Figure 1.1 shows us that the hydrogen atoms are not arranged around the oxygen atom in a straight line. There is an angle of approximately 105° (i.e. a little larger than a right angle) between the hydrogen atoms. The hydrogen atoms have a positive charge, which means that they repulse each other, but at the same time there are two non-bonding electron pairs on the oxygen atom that also repulse the hydrogen atoms. This leads to the molecular structure shown in Figure 1.1. A water molecule can be described as *bipolar*, which means that there is a positive and negative side to the molecule. This polarity is an important property of water as it leads to the bonding between molecules of water: **hydrogen bonding**. The positive side of

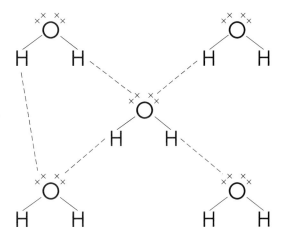

Figure 1.2 The arrangement of water molecules with hydrogen bonds (dotted lines). The stronger covalent bonds between hydrogen and water atoms are shown as solid lines.
Source: Redrawn from McDonald and Kay (1988) and Russell (1976)

the molecule (i.e. the hydrogen side) is attracted to the negative side (i.e. the oxygen atom) of another molecule and a weak hydrogen bond is formed (Figure 1.2). The weakness of this bond means that it can be broken with the application of some energy and the water molecules separate, forming water in a gaseous state (**water vapour**). Although this sounds easy, it actually takes a lot of energy to break the hydrogen bonds between water molecules. This leads to a high specific heat capacity (see p. 4) whereby a lot of energy is absorbed by the water to cause a small rise in energy.

The lack of rigidity in the hydrogen bonds of liquid water gives it two more important properties: a low viscosity and the ability to act as a solvent. The low viscosity comes from water molecules not being so tightly bound together that they cannot separate when a force is applied to them. This makes water an extremely efficient transport mechanism: when a ship applies force to the water molecules they move aside to let it pass! The ability to act as an efficient solvent comes about through water molecules disassociating from each other and being

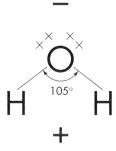

Figure 1.1 The atomic structure of a water molecule. The spare electron pairs on an oxygen atom are shown as small crosses.

able to surround charged compounds contained within them. As described earlier, the ability of water to act as an efficient solvent allows us to use it for washing, the disposal of pollutants, and also allows nutrients to pass from the soil to a plant.

In water's solid state (i.e. ice) the hydrogen bonds become rigid and a three-dimensional crystalline structure forms. What is unusual about water is that the solid form has a lower density than the liquid form, something that is rare in other compounds. This property has profound implications for the world we live in as it means that ice floats on water. More importantly for aquatic life it means that water freezes from the top down rather the other way around. If water froze from the bottom up, then aquatic flora and fauna would be forced upwards as the water froze and eventually end up stranded on the surface of a pond, river or sea. As it is they are able to survive underneath the ice in liquid water. The maximum density of water actually occurs at around 4°C (see Figure 1.3) so that still bodies of water such as lakes and ponds will display thermal stratification, with water close to 4°C sinking downwards.

Water requires a large amount of energy to heat it up. This can be assessed through the **specific heat capacity**, which is the amount of energy required to raise the temperature of a substance by a single degree. Water has a high specific heat capacity relative to other substances: 4.2 kJ/kg/K (kilojoule per kilogram per degree kelvin). This figure means that it requires 4,200 joules of energy to raise the temperature of 1 kilogram of liquid water (approximately 1 litre) by a single degree. In contrast dry soil has a specific heat capacity of around 1.1 kJ/kg/K (it varies according to mineral make up and organic content) and alcohol 0.7 kJ/kg/K. Heating causes the movement of water molecules and that movement requires the breaking of the hydrogen bonds linking them. The large amount of energy required to break the hydrogen bonds in water gives it such a high specific heat capacity.

We can see evidence of water's high specific heat capacity in bathing waters. It is common for sea temperatures to be much lower than air temperatures in high summer as the water is absorbing all of the solar radiation and heating up very slowly. In contrast the water temperature also decreases slowly, leading to the sea often being warmer than the air during autumn and winter. As the water cools down it starts to release the energy that it absorbed as it heated up. Consequently for every drop in temperature of 1°C a single kilogram of water releases 4.2 kJ of energy into the atmosphere. It is this that makes water a climate ameliorator. During the summer months a water body will absorb huge amounts of energy as it slowly warms up; in an area without the water body that energy would heat the earth and consequently air temperatures would be much higher. In the winter the energy is slowly released from the water as it cools down and is available for heating the atmosphere nearby. This is why a maritime climate has cooler summers, but warmer winters, than a continental climate. This is also the mechanism by which large amounts of energy are transported away from the hot equatorial regions towards the cooler poles. As water is evaporated it takes up a large amount of energy that is then released again as the water precipitates (i.e. returns to a liquid form). In the meantime the water has often moved considerable distances in weather systems, taking the latent heat with it. It is estimated that water movement accounts for 70 per cent of lateral global energy transport through latent heat (Mauser and Schädlich, 1998).

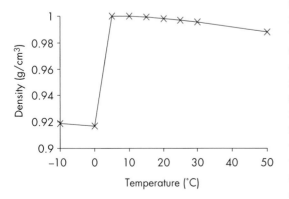

Figure 1.3 The density of water with temperature.

Water acts as a climate ameliorator in one other way: water vapour is a powerful greenhouse gas. Radiation direct from the sun (often referred to as short-wave radiation) passes straight through the atmosphere and may be then absorbed by the earth's surface. This energy is normally re-radiated back from the earth's surface in a different form (often referred to as long-wave radiation). This long-wave radiation is absorbed by the gaseous water molecules and consequently does not escape the atmosphere. This leads to the gradual warming of the earth–atmosphere system as there is an imbalance between the incoming and outgoing radiation. It is the presence of water vapour in our atmosphere (and other gases such as carbon dioxide and methane) that has allowed the planet to be warm enough to support all of the present life forms that exist.

The catchment or river basin

In studying hydrology the most common spatial unit of consideration is the **catchment** or **river basin**. This can be defined as the area of land from which water flows towards a river and then in that river to the sea. The terminology suggests that the area is analogous to a basin where all water moves towards a central point (i.e. the plug hole or in this case the river mouth). The common denominator of any point in a river basin is that wherever rain falls, it will end up in the same place: where the river meets the sea (barring evaporation taking place of course!). A river basin may range in size from a matter of hectares to millions of square kilometres.

A river basin can be defined in terms of its topography through the assumption that all water falling on the surface flows downhill. In this way a catchment boundary can be drawn (as in Figure 1.4) which defines the actual catchment area for a river basin. The assumption that all water flows downhill to the river is not always correct, especially where the underlying geology of a catchment is complicated. It is possible for water to flow as groundwater into another catchment area, creating a problem for the definition of 'catchment area'. These problems aside, the river basin does provide an important

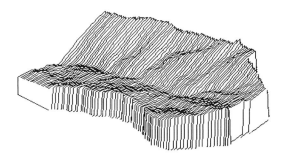

Figure 1.4 A three-dimensional representation of a catchment.

spatial unit for hydrologists to consider how water is moving about and is distributed at a certain time.

THE HYDROLOGICAL CYCLE

As a starting point for the study of hydrology it is useful to consider the **hydrological cycle**. This is a conceptual model of how water moves around between the earth and atmosphere in different states as a gas, liquid or solid. As with any conceptual model it contains many gross simplifications; these are discussed in this section. There are different scales that the hydrological cycle can be viewed at, but it is helpful to start at the large global scale and then move to the smaller hydrological unit of a river basin or catchment.

The global hydrological cycle

Table 1.1 sets out an estimate for the amount of water held on the earth at a single time. These figures are extremely hard to estimate accurately. Estimates cited in Gleick (1993) show a range in total from 1.36 to 1.45 thousand million (or US billion) cubic kilometres of water. The vast majority of this is contained in the oceans and seas. If you were to count groundwater less than 1 km in depth as 'available' and discount snow and ice, then the total percentage of water available for human consumption is around 0.27 per cent. Although this

Table 1.1 Estimated volumes of water held at the earth's surface

	Volume ($\times 10^3$ km^3)	Percentage of total
Oceans and seas	1,338,000	96.54
Ice caps and glaciers	24,064	1.74
Groundwater	23,400	1.69
Permafrost	300	0.022
Lakes	176	0.013
Soil	16.5	0.001
Atmosphere	12.9	0.0009
Marsh/wetlands	11.5	0.0008
Rivers	2.12	0.00015
Biota	1.12	0.00008
Total	1,385,984	100.00

Source: Data from Shiklomanov and Sokolov (1983)

sounds very little it works out at about 146 million litres of water per person per day (assuming a world population of 7 billion); hence the ease with which Stumm (1986) was able to state that there is enough to satisfy all human needs.

Figure 1.5 shows the movement of water around the earth–atmosphere system and is a crude representation of the global hydrological cycle. The cycle consists of **evaporation** of liquid water into water vapour that is moved around the atmosphere. At some stage the water vapour condenses into a liquid (or solid) again and falls to the surface as **precipitation**. The oceans evaporate more water than they receive as precipitation, while the opposite is true over the continents. The difference between precipitation and evaporation in the terrestrial zone is **runoff** (NB This includes the movement of groundwater), which completes the hydrological cycle. As can be seen in Figure 1.5 the vast majority of evaporation and precipitation occurs over the oceans. Ironically this means that the terrestrial zone, which is of greatest concern to hydrologists, is actually rather insignificant in global terms.

The three parts shown in Figure 1.5 (evaporation, precipitation and runoff) are the fundamental processes of concern in hydrology. The figures given in the diagram are global totals but they vary enormously around the globe. With the advent of

satellite monitoring of the earth's surface in the past twenty years it is now possible to gather information on the global distribution of these three processes and hence view how the hydrological cycle varies around the world. In Plates 1 and 2 there are two images of global **rainfall** distribution during 1995, one for January and another for July.

The figure given above of 146 million litres of fresh water per person per year is extremely misleading, as the distribution of available water around the globe varies enormously. The idea of available water considers not only the distribution of rainfall but also population. Table 1.2 gives some indication of those countries that could be considered water rich and water poor. Even this is misleading as a country such as Australia is so large that the high rainfall received in the tropical north-west compensates for the extreme lack of rainfall elsewhere; hence it might be considered water rich. The use of rainfall alone is also misleading as it does not consider the importation of water across borders, through rivers and groundwater movement.

To try and overcome some of the difficulties in interpreting the data in Figure 1.4 and Table 1.2 hydrologists often work at a scale of more relevance to the physical processes occurring. This is frequently the water basin or catchment scale.

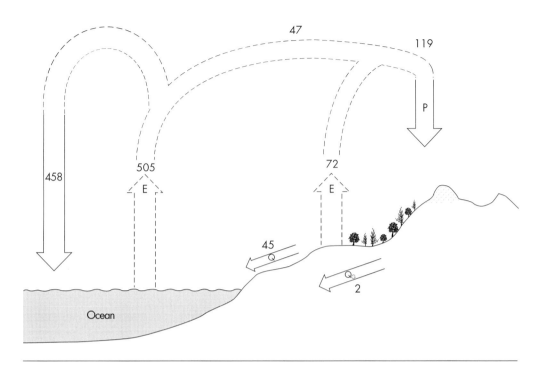

Figure 1.5 The global hydrological cycle. The numbers represent estimates on the total amount of water (thousands of km^3) in each process per annum. E = evaporation; P = precipitation; Q_G = subsurface runoff; Q = surface runoff.
Source: Redrawn from Shiklomanov (1993)

The basin hydrological cycle

At a smaller scale it is possible to view the basin hydrological cycle as a more in-depth conceptual model of processes operating. Figure 1.6 shows an adaptation of the global hydrological cycle to show the processes operating within a catchment. In Figure 1.6 there are still essentially three processes operating (evaporation, precipitation and runoff), but it is possible to subdivide each into different sub-processes. Evaporation is a mixture of open water evaporation (i.e. from rivers and lakes); evaporation from the soil; evaporation from plant surfaces; **interception**; and transpiration from plants. Precipitation can be in the form of **snowfall**, hail, rainfall or some mixture of the three (sleet). Interception of precipitation by plants makes the

water available for evaporation again before it even reaches the soil surface. The broad term 'runoff' incorporates the movement of liquid water above and below the surface of the earth. The movement of water below the surface necessitates an understanding of infiltration into the soil and how the water moves in the unsaturated zone (**throughflow**) and in the saturated zone (**groundwater flow**). All of these processes and sub-processes are dealt with in detail in later chapters; what is important to realise at this stage is that it is part of one continuous cycle that moves water around the globe and that they may all be operating at different times within a river basin.

Table 1.2 Annual renewable water resources per capita (1990 figures) of the seven resource-richest and poorest countries (and other selected countries). Annual renewable water resource is based upon the rainfall within each country; in many cases this is based on estimated figures

Water resource richest countries	Annual internal renewable water resources per capita (thousand m³/yr)	Water resource poorest countries	Annual internal renewable water resources per capita (thousand m³/yr)
Iceland	671.9	Bahrain	0.00
Suriname	496.3	Kuwait	0.00
Guyana	231.7	Qatar	0.06
Papua New Guinea	199.7	Malta	0.07
Solomon Islands	149.0	Yemen Arab Republic	0.12
Gabon	140.1	Saudi Arabia	0.16
New Zealand	117.5	United Arab Emirates	0.19
Canada	109.4	Israel	0.37
Australia	20.5	Kenya	0.59
USA	9.9	United Kingdom	2.11

Source: Data from Gleick (1993)

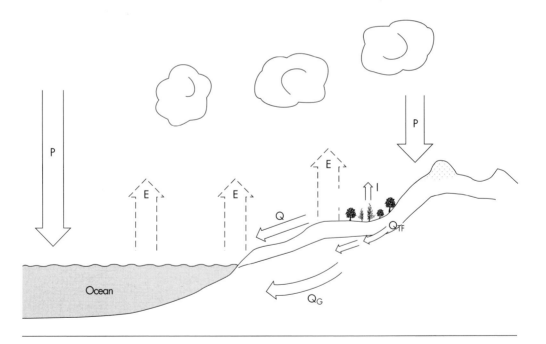

Figure 1.6 Processes in the hydrological cycle operating at the basin or catchment scale. Q = runoff, the subscript G stands for groundwater flow, TF for throughflow; I = interception; E = evaporation; P = precipitation.

THE WATER BALANCE EQUATION

In the previous section it was stated that the hydrological cycle is a conceptual model representing our understanding of which processes are operating within an overall earth–atmosphere system. It is also possible to represent this in the form of an equation, which is normally termed the **water balance equation**. The water balance equation is a mathematical description of the processes operating within a given timeframe and incorporates principles of mass and energy continuity. In this way the hydrological cycle is defined as a closed system whereby there is no mass or energy created nor lost within it. The mass of concern in this case is water.

There are numerous ways of representing the water balance equation but equation 1.1 shows it in its most fundamental form.

$$P \pm E \pm \Delta S \pm Q = 0 \qquad (1.1)$$

where P is precipitation; E is evaporation; ΔS is the change in **storage** and Q is runoff. Runoff is normally given the notation of Q to distinguish it from rainfall which is often given the symbol R and frequently forms the major component of precipitation. The $\pm$ terminology in equation 1.1 represents the fact that each term can be either positive or negative depending on which way you view it – for example, precipitation is a gain (positive) to the earth but a loss (negative) to the atmosphere. As most hydrology is concerned with water on or about the earth's surface it is customary to consider the terms as positive when they represent a gain to the earth.

Two of the more common ways of expressing the water balance are shown in equations 1.2 and 1.3

$$P - E \pm \Delta S - Q = 0 \qquad (1.2)$$

$$Q = P - E \pm \Delta S \qquad (1.3)$$

In both of these the change in storage term can be either positive or negative, as water can be released from storage (positive) or absorbed into storage (negative).

The terms in the water balance equation can be recognised as a series of fluxes and stores. A **flux** is a rate of flow of some quantity (Goudie *et al.*, 1994): in the case of hydrology the quantity is water. The water balance equation assesses the relative flux of water to and from the surface with a storage term also incorporated. A large part of hydrology is involved in measuring or estimating the amount of water involved in this flux transfer and storage of water.

Precipitation in the water balance equation represents the main input of water to a surface (e.g. a catchment). As explained on p. 7, precipitation is a flux of both rainfall and snowfall, while evaporation includes that from open water bodies (lakes, ponds, rivers), the soil surface and vegetation (including transpiration from plants). The storage term includes water in lakes, glaciers, seasonal snow cover and beneath the ground. The runoff flux is also explained on p. 7. In essence it is the movement of liquid water above and below the surface of the earth.

The water balance equation is probably the closest that hydrology comes to having a fundamental theory underlying it as a science, and hence almost all hydrological study is based around it. Field catchment studies are frequently trying to measure the different components of the equation in order to assess others. Nearly all hydrological **models** attempt to solve the equation for a given time span – for example, by knowing the amount of rainfall for a given area and estimating the amount of evaporation and change in storage it is possible to calculate the amount of runoff that might be expected.

Despite its position as a fundamental hydrological theory there is still considerable uncertainty about the application of the water balance equation. It is not an uncertainty about the equation itself but rather about how it may be applied. The problem is that all of the processes occur at a spatial and temporal scale (i.e. they operate over a period of time and within a certain area) that may not coincide with the scale at which we make our measurement or estimation. It is this issue of *scale* that makes hydrology appear an imprecise science and it will be discussed further in the remaining chapters of this book.

OUTLINE OF THE BOOK

The four components of the water balance equation (i.e. precipitation, evaporation, change in storage and in runoff) form the basis of Chapters 2–6. Precipitation is dealt with in Chapter 2, followed by evaporation in Chapter 3. Chapter 4 is a shorter chapter dealing with the interception of precipitation by a canopy. This is a mixture of both evaporation and precipitation, so it has been described separately. Chapter 5 looks at the storage term from the water balance equation, in particular the role of water stored under the earth's surface and as snow and ice. Chapter 6 is concerned with the runoff processes that lead to water flowing down a channel in a stream or river.

Each of Chapters 2–6 starts with a detailed description of the process under review in the chapter. They then move on to contain a section on how it is possible to measure the process, followed by a section on how it may be estimated. In reality it is not always possible to separate between measurement and estimation as many techniques contain an element of both within them, something that is pointed out in various places within these chapters.

Chapter 7 moves away from a description of process and looks at the methods available to analyse streamflow records. This is one of the main tasks within hydrology and three particular techniques are described: hydrograph analysis (including the unit hydrograph), flow duration curves, and frequency analysis. The latter mostly concentrates on **flood frequency analysis**, although there is a short description of how the techniques can be applied to low flows.

Chapter 8 is concerned with water quality in the fresh water environment. The chapter has a description of major water quality parameters, measurement techniques and some strategies used to control water quality.

The final chapter takes an integrated approach to look at different issues of change that affect hydrology. This ranges from water resource management and a changing legislative framework to climate and land use change. These issues are discussed with reference to research studies investigating the different themes. It is intended as a way of capping off the fundamentals of hydrology by looking at real issues facing hydrology in the twenty-first century.

ESSAY QUESTIONS

1   **Discuss the nature of water's physical properties and how important these are in determining the natural climate of the earth.**

2   **Describe how the hydrological cycle varies around the globe.**

3   **How may water-poor countries overcome the lack of water resources within their borders?**

WEBSITES

A warning: although it is often easy to access information via the World Wide Web you should always be careful in utilising it. There is no control on the type of information available or on the data presented. More traditional channels, such as research journals and books, undergo a peer review process where there is some checking of content. This may happen for websites but there is no guarantee that it has happened. You should be wary of treating everything read from the World Wide Web as being correct.

The websites listed here are general sources of hydrological information that may enhance the reading of this book. The majority of addresses are included for the web links provided within their sites. The web addresses are up to date in early 2002 but may change in the future. Hopefully there is enough information provided to enable the use of a search engine to locate updated addresses.

http://www.cig.ensmp.fr/~iahs

International Association of Hydrological Sciences (IAHS): a constituent body of the International Union of Geodesy and Geophysics (IUGG), promoting the interests of hydrology around the world. This has a useful links page.

http://www.cig.ensmp.fr/~hubert/glu/aglo.htm

Part of the IAHS site, this provides a glossary of hydrological terms (in multiple languages).

http://www.worldwater.org

The World's Water, part of the Pacific Institute for Studies in Development, Environment, and Security: this is an organisation that studies water resource issues around the world. There are some useful information sets here.

http://www.uwin.siu.edu

Universities Water Information Network: 'Disseminates information of interest to the water resources community' in the USA.

http://www.catchment.crc.org.au

Cooperative Research Centre for Catchment Hydrology: an Australian research initiative that focuses on tools and information of use in catchment management.

http://www.watsys.sr.unh.edu

Water Systems Analysis Group at the University of New Hampshire: undertakes a diverse group of hydrological research projects at different scales and regions. Much useful information and many useful links.

http://www.hydrology.org.uk

British Hydrological Society: has a links page with mainly UK sites of interest.

http://www.cof.orst.edu/cof/fe/watershed

Hillslope and Watershed Hydrology Team at Oregon State University: this has many good links and information on the latest research.

http://www.nwl.ac.uk/ih

Centre for Ecology and Hydrology (formerly Institute of Hydrology) in the UK: a hydrological research institute. There is a very good worldwide links page here.

http://water.usgs.gov

Water Resources Division of the United States Geological Survey (USGS): provides information on groundwater, surface water and water quality throughout the USA.

http://ghrc.msfc.nasa.gov

Global Hydrology Resource Centre: a NASA site with mainly remote sensing data sets of relevance for global hydrology.

http://daac.gsfc.nasa.gov/campaign-docs/hydrology

The Distributed Active Archive Centre at the Goddard Space Flight Centre: another NASA site with remote sensing data sets of relevance to hydrology.

2

PRECIPITATION

LEARNING OBJECTIVES

When you have finished reading this chapter you should have:

- ■ An understanding of the processes of precipitation formation.
- ■ A knowledge of the techniques for measuring precipitation (rainfall and snow).
- ■ An appreciation of the associated errors in measuring precipitation.
- ■ A knowledge of how to analyse rainfall data spatially and for intensity/duration of a storm.
- ■ A knowledge of some of the methods used to estimate rainfall at the large scale.

PRECIPITATION AS A PROCESS

Precipitation is the release of water from the atmosphere to reach the surface of the earth. The term 'precipitation' covers all forms of water being released by the atmosphere, including snow, hail, sleet and rainfall. It is the major input of water to a river catchment area and as such needs careful assessment in any hydrological study. Although rainfall is relatively straightforward to measure (other forms of precipitation are more difficult) it is notoriously difficult to measure *accurately* and, to compound the problem, is also extremely variable within a catchment area.

Precipitation formation

The ability of air to hold water vapour is temperature dependent: the cooler the air the less water vapour is retained. If a body of warm, moist air is cooled then it will become saturated with water vapour and eventually the water vapour will condense into liquid or solid water (i.e. water or ice droplets). The water will not condense spontaneously however; there need to be minute particles present in the atmosphere, called **condensation nuclei**, upon which the water or ice droplets form. The water or ice droplets that form on condensation nuclei are normally too small to fall to the surface as precipitation; they need to grow in order to have enough mass to overcome uplifting forces within a cloud. So there are three conditions that need to be met prior to precipitation forming:

1 Cooling of the atmosphere
2 Condensation onto nuclei
3 Growth of the water/ice droplets.

Atmospheric cooling

Cooling of the atmosphere may take place through several different mechanisms occurring independently or simultaneously. The most common form of cooling is from the uplift of air through the atmosphere. As air rises the pressure decreases; Boyle's Law states that this will lead to a corresponding cooling in temperature. The cooler temperature leads to less water vapour being retained by the air and conditions becoming favourable for **condensation**. The actual uplift of air may be caused by heating from the earth's surface (leading to **convective precipitation**), an air mass being forced to rise over an obstruction such as a mountain range (this leads to **orographic precipitation**), or from a low pressure weather system where the air is constantly being forced upwards (this leads to **cyclonic precipitation**). Other mechanisms whereby the atmosphere cools include a warm air mass meeting a cooler air mass, and the warm air meeting a cooler object such as the sea or land.

Condensation nuclei

Condensation nuclei are minute particles floating in the atmosphere which provide a surface for the water vapour to condense into liquid water upon. They are commonly less than a micron (i.e. one-millionth of a metre) in diameter. There are many different substances that make condensation nuclei, including small dust particles, sea salts and smoke particles.

Research into generating artificial rainfall has concentrated on the provision of condensation nuclei into clouds, a technique called **cloud seeding**. During the 1950s and 1960s much effort was expended in using silver iodide particles, dropped from planes, to act as condensation nuclei. However, more recent work has suggested that other salts such as potassium chloride are better nuclei. There is much controversy over the worth of cloud seeding. Some studies support its effectiveness (e.g. Gagin and Neumann, 1981; Ben-Zvi, 1988); other authors query the results (e.g. Rangno and Hobbs, 1995), while others suggest that it only works in certain atmospheric conditions and with certain cloud types (e.g. Changnon *et al.*, 1995). More recent work in South Africa has concentrated on using hygroscopic flares to release chloride salts into the base of convective storms, with some success (Mather *et al.*, 1997). Interestingly, this approach was first noticed through the discovery of extra heavy rainfall occurring over a paper mill in South Africa that was emitting potassium chloride from its chimney stack (Mather, 1991).

Water droplet growth

Water or ice droplets formed around condensation nuclei are normally too small to fall directly to the ground; that is, the forces from the upward draught within a cloud are greater than the gravitational forces pulling the microscopic droplet downwards. In order to overcome the upward draughts it is necessary for the droplets to grow from an initial size of 1 micron to around 3,000 microns (3 mm). The vapour pressure difference between a droplet and the surrounding air will cause it to grow through condensation, albeit rather slowly. When the water droplet is ice the vapour pressure difference with the surrounding air becomes greater and the water vapour sublimates onto the ice droplet. This will create a precipitation droplet faster than condensation onto a water droplet, but is still a slow process. The main mechanism by which raindrops grow within a cloud is through *collision and coalescence*. Two raindrops collide and join together (coalesce) to form a larger droplet that may then collide with many more before falling towards the surface as rainfall or another form of precipitation.

Another mechanism leading to increased water droplet size is the so-called **Bergeron process**. The pressure exerted within the parcel of air, by having the water vapour present within it, is called the **vapour pressure**. The more water vapour present the greater the vapour pressure. Because there is a maximum amount of water vapour that can be held by the parcel of air there is also a maximum vapour pressure, the so-called **saturation vapour pressure**. The saturation vapour pressure is greater over a water droplet than an ice droplet because it is easier for water molecules to escape from the surface of a liquid than a solid. This creates a water vapour gradient between water droplets and ice crystals so that water vapour moves from the water droplets to the ice crystals, thereby increasing the size of the ice crystals. Because clouds are usually a mixture of water vapour, water droplets and ice crystals, the Bergeron process may be a significant factor in making water droplets large enough to become rain drops (or ice/snow crystals) that overcome gravity and fall out of the clouds.

The mechanisms of droplet formation within a cloud are not completely understood. The relative proportion of condensation-formed, collision-formed, and Bergeron-process-formed droplets depends very much on the individual cloud circumstances and can vary considerably. As a droplet is moved around a cloud it may freeze and thaw several times, leading to different types of precipitation (see Table 2.1).

PRECIPITATION DISTRIBUTION

The amount of precipitation falling over a location varies both spatially and temporally (with time).

The different influences on the precipitation can be divided into static and dynamic influences. Static influences are those such as altitude, aspect and slope; they do not vary between storm events. Dynamic influences are those that do change and are by and large caused by variations in the weather. At the global scale the influences on precipitation distribution are mainly caused by differing weather patterns, but there are static factors such as topography that can cause major variations through a **rain shadow effect** (see pp. 16–17). At the continental scale large differences in rainfall can be attributed to a mixture of static and dynamic factors. In Figure 2.1 the rainfall distribution across the USA shows marked variations. Although mountainous areas have a higher rainfall, and also act as a block to rainfall reaching the drier centre of the country, they do not provide the only explanation for the variations evident in Figure 2.1. The higher rainfall in the north-west states (Oregon and Washington) is linked to wetter cyclonic weather systems from the northern Pacific that do not reach down to southern California. Higher rainfall in Florida and other southern states is linked to the warm waters of the Caribbean sea. These are examples of dynamic influences as they vary between rainfall events.

At smaller scales the static factors are often more dominant, although it is not uncommon for quite large variations in rainfall across a small area caused by individual storm clouds to exist. An example of this was seen on 3 July 2000 when an intense rainfall event caused flooding in the village of Epping Green, Essex, UK. Approximately 10 mm of rain fell within one hour, although there was no recorded rainfall in the village of Theydon

Table 2.1 Classes of precipitation used by the UK Meteorological Office

Class	Definition
Rain	Liquid water droplets between 0.5 and 7 mm in diameter
Drizzle	A subset of rain with droplets less than 0.5 mm
Sleet	Freezing raindrops; a combination of snow and rain
Snow	Complex ice crystals agglomerated
Hail	Balls of ice between 5 and 125 mm in diameter

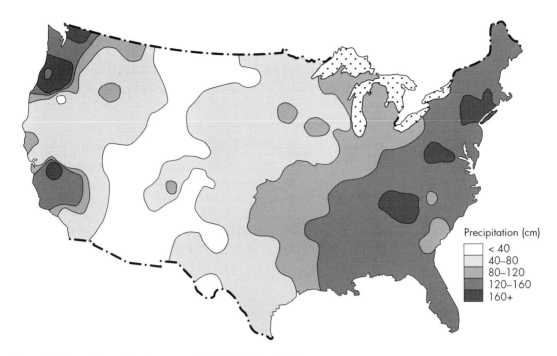

Figure 2.1 Annual precipitation across the USA during 1996.
Source: Redrawn with data from the National Atmospheric Deposition Program

Bois approximately 10 km to the south. This large spatial difference in rainfall was caused by the scale of the weather system causing the storm – in this case a convective thunderstorm. Often these types of variation lessen in importance over a longer time-scale so that the annual rainfall in Epping Green and Theydon Bois is very similar, whereas the daily rainfall may differ considerably. For the hydrologist, who is interested in rainfall at the small scale, the only way to try and characterise these dynamic variations is through having as many **rain gauges** as possible within a study area.

Static influences on precipitation distribution

It is easier for the hydrologist to account for static variables such as those discussed below.

Altitude

It has already been explained that temperature is a critical factor in controlling the amount of water vapour that can be held by air. The cooler the air is, the less water vapour can be held. As temperature decreases with altitude it is reasonable to assume that as an air parcel gains altitude it is more likely to release the water vapour and cause higher rainfall. This is exactly what does happen and there is a strong correlation between altitude and rainfall: so-called *orographic precipitation*.

Aspect

The influence of aspect is less important than altitude but it may still play an important part in the distribution of precipitation throughout a catchment. In the humid mid-latitudes (35° to 65° north or south of the equator) the predominant source of rainfall is through cyclonic weather systems arriving from the west. Slopes within a catchment that face eastwards will naturally be more sheltered from the rain than those facing westwards. The same principle applies everywhere: slopes with aspects facing away from the predominant weather patterns will receive less rainfall than their opposites.

Slope

The influence of slope is only relevant at a very small scale. Unfortunately the measurement of rainfall occurs at a very small scale (i.e. a rain gauge). The difference between a rain gauge on a hillslope that is level, compared to one parallel to the slope, may be significant. It is possible to calculate this difference if it is assumed that rain falls vertically – but of course rain does not always fall vertically. Consequently the effect of slope on rainfall measurements is normally ignored.

Rain shadow effect

Where there is a large and high land mass it is common to find the rainfall considerably higher on one side than the other. This is through a combination of altitude, slope, aspect and dynamic weather direction influences and can occur at many different scales (see Case Study below).

Case study

THE RAIN SHADOW EFFECT

The predominant weather pattern for the South Island of New Zealand is a series of rain-bearing depressions sweeping up from the Southern Ocean, interrupted by drier blocking anticyclones. The Southern Alps form a major barrier to the fast-moving depressions and have a huge influence on the rainfall distribution within the South Island. Formed as part of tectonic uplift along the Pacific/Indian plate boundary, the Southern Alps stretch the full length of the South Island (approximately 700 km) and at their highest point are over 3,000 m above mean sea level.

The predominant weather pattern has a westerly airflow, bringing moist air from the Tasman Sea onto the South Island. As this air is forced up over the Southern Alps it cools down and releases much of its moisture as rain and snow. As the air descends on the eastern side of the mountains it warms up and becomes a föhn wind, referred to locally as a 'nor-wester'. The annual rainfall patterns for selected stations in the South Island are shown in Figure 2.2. The rain shadow effect can be clearly seen with the west coast rainfall being at least four times that of the east. Table 2.2 also illustrates the point, with the number of rain days at different sites in a cross section across the South Island.

This pattern of rain shadow is seen at many different locations around the globe. It does not require as large a barrier as the Southern Alps – anywhere with a significant topographical barrier is likely to cause some form of rain shadow. Hayward and Clarke (1996) present data showing a strong rain shadow across the Freetown Peninsula in Sierra Leone. They analysed mean monthly rainfall in thirty-one gauges within a 20 × 50 km area, and found that the rain shadow effect was most marked during the monsoon months of June to October. The gauges in locations facing the ocean (south-west aspect) caught considerably more rainfall during the monsoon than those whose aspect was towards the north-east and behind a small range of hills.

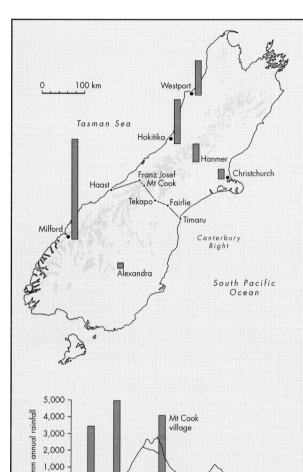

Figure 2.2 Rainfall distribution across the Southern Alps of New Zealand (South Island). Shaded areas on the map are greater than 1,500 m in elevation. A clear rain shadow effect can be seen between the much wetter west coast and the drier east.

Table 2.2 Average annual rainfall and rain days for a cross section across the South Island

Weather station	Height above mean sea level	Annual rainfall (mm)	Rain days
Haast	30	5,840	175
Mt Cook village	770	4,293	161
Tekapo	762	604	77
Timaru	25	541	75

Note: More details on weather differentials across the South Island of New Zealand are in Sinclair *et al.* (1996)
Source: Data from New Zealand Met. Service and other miscellaneous sources

MEASUREMENT

For hydrological analysis it is important to know how much precipitation has fallen and when this occurred. The usual expression of precipitation is as a vertical depth of liquid water. Rainfall is measured by millimetres or inches depth, rather than by volume such as litres or cubic metres. The measurement is the depth of water that would accumulate on the surface if all the rain remained where it had fallen (Shaw, 1994). Snowfall may also be expressed as a depth, although for hydrological purposes it is most usefully described in water equivalent depth (i.e. the depth of water that would be present if the snow melted). This is a recognition that snow takes up a greater volume (as much as 90 per cent more) for the same amount of liquid water.

There is a strong argument that can be made to say that there is no such thing as precipitation measurement at the catchment scale as it varies so tremendously over a small area. The logical endpoint to this argument is that all measurement techniques are in fact precipitation estimation techniques. For the sake of clarity in this text precipitation measurement techniques refer to the methods used to quantify the volume of water present, as opposed to estimation techniques where another variable is used as a surrogate for the water volume.

Rainfall measurement

The instrument for measuring rainfall is called a *rain gauge*. A rain gauge measures the volume of water that falls onto a horizontal surface delineated by the rain gauge rim (see Figure 2.3). The volume is converted into a rainfall depth through division by the rain gauge surface area. The design of a rain gauge is not as simple as it may seem at first glance as there are many errors and inaccuracies that need to be minimised or eliminated.

There is a considerable scientific literature studying the accuracy and errors involved in measuring rainfall. It needs to be borne in mind that a rain gauge represents a very small point measurement (or sample) from a much larger area that is covered

Figure 2.3 A rain gauge sitting above the surface to avoid splash.

by the rainfall. Any errors in measurement will be amplified hugely because the rain gauge collection area represents such a small sample size. Because of this amplification it is extremely important that the design of a rain gauge negates any errors and inaccuracies.

The four main sources of error in measuring rainfall that need consideration in designing a method for the accurate measurement of rainfall are:

1 Losses due to evaporation
2 Losses due to wetting of the gauge
3 Over-measurement due to splash from surrounding area
4 Under-measurement due to turbulence around the gauge.

Worked example of rain gauge error magnification

The standard UK design rain gauge has a 127 mm diameter rim which gives a surface area of 0.013 m². If we were to assume that there was only one rain gauge in a small catchment of 10 km² then the ratio of sample to population size is 1:76,923,077. From this it is possible to calculate the size of any potential errors.

If the rain gauge measured 1 ml (1 cm³) less rainfall than reality then a water balance analysis would conclude that there is 769.23 m³ less water in the catchment than in reality. It is very easy for rainfall to be accurate to less than 1 ml, especially for a poorly maintained, manually recorded instrument. This kind of error magnification has important implications for water resource management where it is important to have a good knowledge of how much water is in a catchment system.

The example given is a conservative estimate. In a heavily instrumented country such as the UK there is an average rain gauge density of approximately one per 30 km². In parts of the world such as Africa the density may be as low as one gauge per 2000 km². In these cases the error magnification between sample size and total catchment area is hugely increased.

Evaporation losses

A rain gauge can be any collector of rainfall with a known collection area; however, it is important that any rainfall that does collect is not lost again through evaporation. In order to eliminate, or at least lessen, this loss rain gauges are funnel shaped. In this way the rainfall is collected over a reasonably large area and then any water collected is passed through a narrow aperture to a collection tank underneath. Because the collection tank has a narrow top (i.e. the funnel mouth) there is very little interchange of air with the atmosphere above the gauge. As will be explained in Chapter 3, one of the necessary requirements for evaporation is the turbulent mixing of saturated air with drier air above. By restricting this turbulent transfer there is little evaporation that can take place. In addition to this, the water awaiting measurement is kept out of direct sunlight so that it will not be warmed; hence there is a low evaporation loss.

Wetting loss

As the water trickles down the funnel it is inevitable that some water will stay on the surface of the funnel and can be lost to evaporation or not measured in the collection tank. This is often referred to as a *wetting loss*. These losses will not be large but may be significant, particularly if the rain is falling as a series of small events on a warm day. In order to lessen this loss it is necessary to have steep sides on the funnel and to have a non-stick surface. The standard UK Meteorological Office rain gauge is made of copper to create a non-stick surface, although many modern rain gauges are made of non-adhesive plastics.

Rain splash

The perfect rain gauge should measure the amount of rainfall that would have fallen on a surface if the gauge was not there. This suggests that the ideal situation for a rain gauge is flush with the surface. A difficulty arises, however, as a surface-level gauge is likely to over-measure the catch due to rain landing adjacent to the gauge and splashing into it. If there was an equal amount of splash going out of the gauge then the problem might not be so severe, but the sloping sides of the funnel (to reduce evaporative losses) mean that there will be very little splash-out. In extreme situations it is even possible that the rain gauge could be flooded by water flowing over the surface or covered by snow. To overcome the splash, flooding and snow coverage problem the rain gauge can be raised up above the ground or placed in the middle of a non-splash grid (see Figure 2.6).

Turbulence around a raised gauge

If a rain gauge is raised up above the ground (to reduce splash) another problem is created due to air turbulence around the gauge. The rain gauge presents an obstacle to the wind and the consequent aerodynamic interference leads to a reduced catch (see Figure 2.4). The amount of loss is dependent on both the wind speed and the raindrop diameter (Nešpor and Sevruk, 1999). At wind speeds of 20 km/hr (Beaufort scale 2) the loss could be up to 20 per cent, and in severe winds of 90 km/hr (Beaufort scale 8) up to 40 per cent (Bruce and Clark, 1980; Rodda and Smith, 1986). The higher a gauge is from the surface the greater the loss of accuracy. This creates a major problem for gauges in areas that receive large snowfalls as they need to be raised to avoid surface coverage.

One method of addressing these turbulence difficulties is through the fitting of a shield to the rain gauge (see Figure 2.5). A rain gauge shield can take many forms but is often a series of batons surrounding the gauge at its top height. The shield acts as a calming measure for wind around the gauge and has been shown to greatly improve rain gauge accuracy.

The optimum rain gauge design

There is no perfect rain gauge. The design of the best gauge for a site will be influenced by the individual conditions at the site (e.g. prevalence of

Figure 2.5 Baffles surrounding a rain gauge to lessen the impact of wind turbulence. The gauge is above ground because of snow cover during the winter.

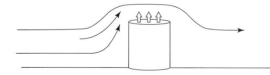

Figure 2.4 The effect of wind turbulence on a raised rain gauge. An area of reduced pressure (and uplift) develops above the gauge in a similar manner to an aircraft wing. This reduces the rain gauge catch.

snowfall, exposure, etc.). A rain gauge with a non-splash surround, such as in Figure 2.6, can give very accurate measurement but it is prone to coverage by heavy snowfall so cannot always be used. The non-splash surround allows adjacent rainfall to pass through (negating splash) but acts as an extended soil surface for the wind, thereby eliminating the turbulence problem from raised gauges. This may be the closest that it is possible to get to measuring the amount of rainfall that would have fallen on a surface if the rain gauge was not there.

The standard UK Meteorological Office rain gauge has been adopted around the world (although not everywhere) as a compromise between the factors influencing rain gauge accuracy. It is a brass rimmed rain gauge of 5 inches (127 mm) diameter standing 1 foot (305 mm) above the ground. The lack of height above ground level is a reflection

Figure 2.6 Surface rain gauge with non-splash surround.

of the low incidence of snowfall in the UK; in countries such as Russia and Canada, where winter snowfall is the norm, gauges may be raised as high as 2 m above the surface. There is general recognition that the UK standard rain gauge is not the best design for hydrology, but it does represent a reasonable compromise. There is a strong argument to be made against changing its design. Any change in the measurement instrument would make an analysis of past rainfall patterns difficult due to the differing accuracy.

Siting of a rain gauge

Once the best measurement device has been chosen for a location there is still a considerable measure-

ment error that can occur through incorrect siting. The major problem of rain gauge siting in hydrology is that the scientist is trying to measure the rainfall at a location that is representative of a far greater area. It is extremely important that the measurement location is an appropriate surrogate for the larger area. If the area of interest is a forested catchment then it is reasonable to place your rain gauge beneath the forest canopy; likewise, within an urban environment it is reasonable to expect interference from buildings because this is what is happening over the larger area. What is extremely important is that there are enough rain gauges to try and quantify the spatial and temporal variations.

The rule-of-thumb method for siting a rain gauge is that the angle when drawn from the top of the rain gauge to the top of the obstacle is less than 30° (see Figure 2.7). This can be approximated as at least twice the height of the obstacle away from the gauge. Care needs to be taken to allow for the future growth of trees so that at all times during the rainfall record the distance apart is at least twice the height of an obstacle.

Gauges for the continuous measurement of rainfall

The standard UK Meteorological Office rain gauge collects water beneath its funnel and this volume is read once a day. Often in hydrology the data needs to be measured at a finer timescale than this,

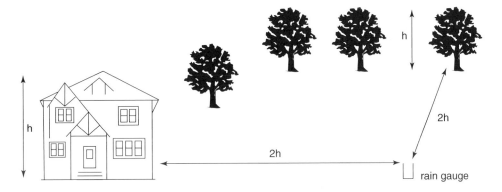

Figure 2.7 Siting of a rain gauge away from obstructions.

particularly in the case of individual storms which often last much less than a day. The most common modern method for measuring continuous rainfall uses a tipping-bucket rain gauge. These are very simple devices that can be installed relatively cheaply, although they do require a data-logging device nearby. The principle behind the tipping-bucket rain gauge is that as the rain falls it fills up a small 'bucket' that is attached to another 'bucket' on a balanced cross arm (see Figure 2.8). The 'buckets' are very small plastic containers at the end of each cross arm. When the bucket is full it tips the balance so that the full bucket is lowered down and empties out. At the time of tipping a magnet attached to the balance arm closes a small reed switch which sends an electrical signal to a data-logging device. This then records the exact time of the tipped bucket. If the rain continues to fall it fills the bucket on the other end of the cross arm until it too tips the balance arm, sending another electrical impulse to the data logger. In this way a near continuous measurement of rainfall with time can be obtained.

It is important that the correct size of tipping bucket is used for the prevailing conditions. If the

buckets are too small then a very heavy rainfall event will cause them to fill too quickly and water be lost through overspill while the mechanism tips. If the buckets are too large then a small rainfall event may not cause the cross arm to tip and the subsequent rainfall event will appear larger than it actually was. The tipping-bucket rain gauge shown in Figure 2.8 has an equivalent depth of 0.2 mm of rain which works well for field studies in south-east England.

Snowfall measurement

The measurement of snowfall has similar problems to those presented by rainfall, but they are often more extreme. There are two methods used for measuring snowfall: using a gauge like a rain gauge; or depth that is present on the ground. Both of these methods have very large errors associated with them, predominantly caused by the way that snow falls through the atmosphere and is deposited on the gauge or ground. Most, although not all, snowflakes are more easily transported by the wind than raindrops. When the snow reaches the ground it is easily blown around in a secondary manner (drifting). This can be contrasted to liquid water where, upon reaching the ground, it is either absorbed by the soil or moves across the surface. Rainfall is very rarely picked up by the wind again and redistributed in the manner that drifting snow is. For the snow gauge this presents problems that are analogous to rain splash. For the depth gauge the problem is due to uneven distribution of the snow surface: it is likely to be deeper in certain situations.

Rain gauge modification to include snowfall

One modification that needs to be made to a standard rain gauge in order to collect snowfall is a heated rim so that any snow falling on the gauge melts to be collected as liquid water. Failure to have a heated rim may mean that the snow builds up on the gauge surface until it overflows. Providing a heated rim is no simple logistic exercise as it necessitates a power source (difficult in remote areas)

Figure 2.8 The insides of a tipping-bucket rain gauge. (NB This is the same gauge as in Figure 2.3). The 'buckets' are the small white, triangular reservoirs. These are balanced and when full they tip over bringing the black arm past the other stationary arm. In doing so a small electrical current is passed to a data logger.

and the removal of collected water well away from the heat source to minimise evaporation losses.

A second modification is to raise the gauge well above ground level so that as snow builds up the gauge is still above this surface. Unfortunately the raising of the gauge leads to an increase in the turbulence errors described for rain gauges. For this reason it is normal to have wind deflectors or shields surrounding the gauge.

Snow depth

The simplest method of measuring snow depth is the use of a core sampler. This takes a core of snow, recording its depth at the same time, that can then be melted to derive the water equivalent depth. It is this that is of importance to a hydrologist. The major difficulties of a core sample are that it is a non-continuous reading (similar to daily rainfall measurement), and the position of coring may be critical (because of snow drifting).

A second method of measuring snow depth is to use a **snow pillow**. This is described further on pp. 63–66.

MOVING FROM POINT MEASUREMENT TO AREAL ESTIMATION

The measurement techniques described here have all concentrated on measuring rainfall at a precise location (or at least over an extremely small area – see the earlier Worked Example on p. 19). In reality the hydrologist needs to know how much precipitation has fallen over a far larger area, usually a catchment. To move from point measurements to an areal estimation it is necessary to employ some form of spatial averaging. The spatial averaging must attempt to account for an uneven spread of rain gauges in the catchment and the various factors that we know influence rainfall distribution (e.g. altitude, aspect and slope). A simple arithmetic mean will only work where a catchment is sampled by uniformly spaced rain gauges and where there

is no diversity in topography. If these conditions were ever truly met then it is unlikely that there would be more than one rain gauge sampling the area. Hence it is very rare to use a simple averaging technique.

There are different statistical techniques that address the spatial distribution issues, and with the growth in use of **Geographic Information Systems (GIS)** it is often a relatively trivial matter to do the calculation. As with any computational task it is important to have a good knowledge of how the technique works so that any shortcomings are fully understood. Three techniques are described here: **Thiessen's polygons**, the **hypsometric method**, and the **isohyetal method**. These methods are explored further in a Case Study on p. 26.

Thiessen's polygons

Thiessen was an American engineer working around the start of the twentieth century who devised a simple method of overcoming an uneven distribution of rain gauges within a catchment (very much the norm). Essentially Thiessen's polygons attach a representative area to each rain gauge. The size of the representative area (a polygon) is based on how close each gauge is to the others surrounding it.

Each polygon is drawn on a map; the boundaries of the polygons are equidistant from each gauge and drawn at a right angle (orthogonal) to an imaginary line between two gauges (see Figure 2.9). Once the polygons have been drawn the area of each polygon surrounding a rain gauge is found. The areal rainfall (R) is calculated using the following formula:

$$R = \sum_{i=1}^{n} \frac{r_i a_i}{A}$$

where r_i is the rainfall at gauge i, a_i is the area of the polygon surrounding rain gauge i, and A is the total catchment area.

The **areal rainfall** value using Thiessen's polygons is a weighted mean, with the weighting being based upon the size of each representative area (polygon). This technique is only truly valid where

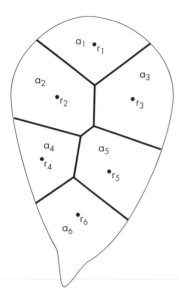

Figure 2.9 Thiessen's polygons for a series of rain gauges (r_i) within an imaginary catchment. The area of each polygon is denoted as a_i. Locations of rain gauges are indicated by bullets.

the topography is uniform within each polygon so that it can be safely assumed that the rainfall distribution is uniform within the polygon. This would suggest that it can only work where the rain gauges are located initially with this technique in mind (i.e. *a priori*).

Hypsometric method

As it is well known that rainfall is positively influenced by altitude (i.e. the higher the altitude the greater the rainfall) it is reasonable to assume that knowledge of the catchment elevation can be brought to bear on the areal rainfall estimation problem. The simplest indicator of the catchment elevation is the hypsometric (or hypsographic) curve. This is a graph showing the proportion of a catchment above or below a certain elevation. The values for the curve can be derived from maps using a planimeter or using a digital elevation model (DEM) in a GIS.

The hypsometric method of calculating areal rainfall then calculates a weighted average based on the proportion of the catchment between two elevations and the measured rainfall between those elevations:

$$R = \sum_{j=1}^{m} r_j p_j$$

where r_j is the average rainfall between two contour intervals and p_j is the proportion of the total catchment area between those contours (derived from the hypsometric curve). The r_j value may be an average of several rain gauges where there is more than one at a certain contour interval. This is illustrated in Figure 2.10 where the shaded area (a_3) has two gauges within it. In this case the p_j will be an average of r_4 and r_5.

Intuitively this is producing representative areas for one or more gauges based on contours and spacing, rather than just on the latter for Thiessen's

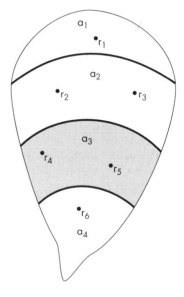

Figure 2.10 Calculation of areal rainfall using the hypsometric method. The shaded region is between two contours. In this case the rainfall is an average between the two gauges within the shaded area. Locations of rain gauges are indicated by bullets.

polygons. There is an inherent assumption that elevation is the only topographic parameter affecting rainfall distribution (i.e. slope and aspect are ignored). It also assumes that the relationship between altitude and rainfall is linear, which is not always the case and warrants exploration before using this technique.

Isohyetal and other smoothed surface techniques

Where there is a large number of gauges within a catchment the most obvious weighting to apply on a mean is based on measured rainfall distribution rather than on surrogate measures as described above. In this case a map of the catchment rainfall distribution can be drawn by interpolating between the rainfall values, creating a smoothed rainfall surface. The traditional isohyetal method involved drawing isohyets (lines of equal rainfall) on the map and calculating the area between each isohyet. The areal average could then be calculated as

$$R = \sum_{i=1}^{n} \frac{r_i a_i}{A}$$

where a_i is the area between each isohyet and r_i is the average rainfall between the isohyets. This technique is analogous to Figure 2.10, except in this case the contours will be of rainfall rather than elevation.

With the advent of GIS the interpolating and drawing of isohyets can be done relatively easily, although there are several different ways of carrying out the interpolation. The interpolation subdivides the catchment into small grid cells and then assigns a rainfall value for each grid cell (this is the smoothed rainfall surface). The simplest method of interpolation is to use a nearest neighbour analysis, where the assigned rainfall value for a grid square is proportional to the nearest rain gauges. A more complicated technique is to use **kriging**, where the interpolated value for each cell is derived with knowledge on how closely related the nearby gauges are to each other in terms of their co-variance.

A fuller explanation of these techniques is provided by Bailey and Gatrell (1995).

An additional piece of information that can be gained from interpolated rainfall surfaces is the likely rainfall at a particular point within the catchment. This may be more useful information than total rainfall over an area, particularly when needed for numerical simulation of hydrological processes.

The difficulty in moving from the point measurement to an areal average is a prime example of the problem of scale that besets hydrology. The scale of measurement (i.e. the rain gauge surface area) is far smaller than the catchment area that is frequently our concern. Is it feasible to simply scale up our measurement from point sources to the overall catchment? Or should there be some form of scaling factor to acknowledge the large discrepancy? There is no easy answer to these questions and they are the type of problem that research in hydrology will be investigating in the twenty-first century.

RAINFALL INTENSITY AND STORM DURATION

Water depth is not the only rainfall measure of interest in hydrology; also of importance is the **rainfall intensity** and **storm duration**. These are simple to obtain from an analysis of rainfall records using frequency analysis. The rainfall needs to be recorded at a short time interval (i.e. an hour or less) to provide meaningful data.

Figure 2.12 shows the rainfall intensity for a rain gauge at Bradwell-on-Sea, Essex, UK. It is evident from the diagram that the majority of rain falls at very low intensity: 0.4 mm per hour is considered as light rain. This may be misleading as the rain gauge recorded rainfall every hour and the small amount of rain may have fallen during a shorter period than an hour i.e. a higher intensity but lasting for less than an hour. During the period of measurement there were recorded rainfall intensities greater than 4.4 mm/hr (maximum 6.8 mm/hr) but they were so few as to not show up on the histogram scale used in Figure 2.12. This may be a reflection of only two

Case study

RAINFALL DISTRIBUTION IN A SMALL STUDY CATCHMENT

It is well known that large variations in rainfall occur over quite a small spatial scale. Despite this, there are not many studies that have looked at this problem in detail. One study that has investigated spatial variability in rainfall was carried out in the Plynlimon research catchments in mid-Wales (Clarke et al., 1973). In setting up a hydrological monitoring network in the Wye and Severn catchments thirty-eight rain gauges were installed to try and characterise the rainfall variation. The rainfall network had eighteen rain gauges in the Severn catchment (total area 8.7 km²) and twenty gauges in the Wye (10.55 km²).

The monthly data for a period between April 1971 and March 1973 were analysed to calculate areal average rainfall using contrasting methods. The results from this can be seen in Figure 2.11. The most startling feature of Figure 2.11 is the lack of difference in calculated values and that they

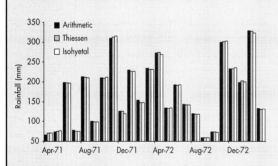

Figure 2.11 Areal mean rainfall (monthly) for the Wye catchment, calculated using three different methods.
Source: Data from Clarke et al. (1973)

follow no regular pattern. At times the arithmetic mean is greater than the other while in other months it is less. When the total rainfall for the two-year period is looked at, the Thiessen's calculation is 0.3 per cent less than the arithmetic mean, while the isohyetal method is 0.4 per cent less.

When the data were analysed to see how many rain gauges would be required to characterise the rainfall distribution fully it was found that the number varied with the time period of rainfall and the season being measured. When monthly data were looked at there was more variability in winter rainfall than summer. For both winter and summer it showed that anything less than five rain gauges (for the Wye) increased the variance markedly.

A more detailed statistical analysis of hourly mean rainfall showed a far greater number of gauges were required. Four gauges would give an accuracy in areal estimate of around 50 per cent, while a 90 per cent accuracy would require 100 gauges (Clarke et al., 1973: 62).

The conclusions that can be drawn from the study of Clarke et al. (1973) are of great concern to hydrology. It would appear that even for a small catchment a large number of rain gauges are required to try and estimate rainfall values properly. This confirms the statement made at the start of this chapter: although rainfall is relatively straightforward to measure it is notoriously difficult to measure accurately and, to compound the problem, is also extremely variable within a catchment area.

years of records being analysed, which introduces an extremely important concept in hydrology: the **frequency–magnitude** relationship. With rainfall (and runoff – see Chapters 6 and 7) the larger the

rainfall event the less frequent we would expect it to be. This is not a linear relationship; as illustrated in Figure 2.12 the curve declines in a non-linear fashion. If we think of the relative frequency as a

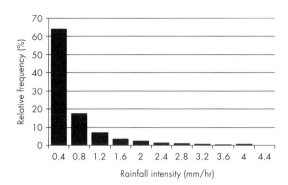

Figure 2.12 Rainfall intensity curve for Bradwell-on-Sea, Essex, UK. Data are hourly recorded rainfall from April 1995 to April 1997.

probability then we can say that the chances of having a low rainfall event are very high: a low magnitude–high frequency event. Conversely the chances of having a rainfall intensity greater than 5 mm/hr are very low (but not impossible): a high magnitude–low frequency event.

In Figure 2.13 the storm duration records for two different sites are compared. The Bradwell-on-Sea site has a huge majority of its rain events lasting one hour or less. In contrast the Ahoskie site has only 20 per cent of its storms lasting one hour or less

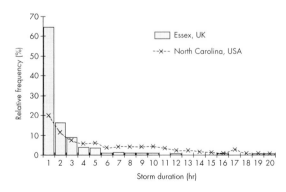

Figure 2.13 Storm duration curves. The bars are for the same data set as Figure 2.12 and the dashed line for Ahoskie, North Carolina.
Source: Ahoskie data are redrawn from Wanielista (1990)

but many more than Bradwell-on-Sea that last four hours or more. When the UK site rainfall and intensity curves are looked at together (i.e. Figures 2.12 and 2.13) it can be stated that Bradwell-on-Sea experiences a predominance of low intensity, short duration rainfall events and very few long duration, high intensity storms. This type of information is extremely useful to a hydrologist investigating the likely runoff response that might be expected for the rainfall regime.

SURROGATE MEASURES FOR ESTIMATING RAINFALL

The difficulties in calculation an areal precipitation value from point measurements make the estimation of areal precipitation an attractive proposition. There are two techniques that make some claim to achieving this: radar, and **satellite remote sensing**. These approaches have many similarities, but they differ fundamentally in the direction of measurement. Radar looks from the earth up into the atmosphere and tries to estimate the amount of precipitation falling over an area. Satellite remote sensing looks from space down towards the earth surface and attempts to estimate the amount of precipitation falling over an area.

Radar

The main use of ground-based radar is in weather forecasting where it is used to track the movement of rain clouds and fronts across the earth's surface. This in itself is interesting but does not provide the hydrological requirement of estimating how much rain is falling over an area.

There are several techniques used for radar, although they are all based on similar principles. Radar is an acronym (*ra*dio *d*etection *a*nd *r*anging). A wave of electromagnetic energy is emitted from a unit on the ground and the amount of wave reflection and return time is recorded. The more water there is in a cloud the more electromagnetic energy is reflected back to the ground and detected by the radar unit. The quicker the reflected wave reaches back to ground the closer the cloud is to the surface. The most difficult part of this technique is in finding the best wavelength of electromagnetic radiation to emit and detect. It is important that the electromagnetic wave is reflected by liquid water in the cloud, but not atmospheric gases and/or changing densities of the atmosphere. A considerable amount of research effort has gone into trying to find the best wavelengths for ground-based radar to use. The solution appears to be that it is somewhere in the microwave band (commonly c-band), but that the exact wavelength depends on the individual situation being studied (Cluckie and Collier, 1991).

Studies have shown a good correlation between reflected electromagnetic waves and rainfall intensity. Therefore, this can be thought of as a surrogate measure for estimating rainfall. If an accurate estimate of rainfall intensity is required then a relationship has to be derived using several calibrating rain gauges. Herein lies a major problem: with this type of technique there is no universal relationship that can be used to derive rainfall intensity from cloud reflectivity. An individual calibration has to be derived for each site and this may involve several years of measuring point rainfall coincidentally with cloud reflectivity. This is not a cheap option and the cost prohibits its widespread usage, particularly in areas with poor rain gauge coverage.

In Britain the UK Meteorological Office operates a series of fifteen weather radar with a 5 km resolution that provide images every 15 minutes. This is a more intensive coverage than could be expected in most countries. Although portable radar can be used for rainfall estimation, their usage has been limited by the high cost of purchase.

Satellite remote sensing

The atmosphere-down approach of satellite remote sensing is quite different from the ground-up approach of radar – fundamentally because the sensor is looking at the top of a cloud rather than

the bottom. It is well established that a cloud most likely to produce rain has an extremely bright and cold top. These are the characteristics that can be observed from space by a satellite sensor. The most common form of satellite sensor is passive (this means it receives radiation from another source, normally the sun, rather than emitting any itself the way radar does) and detects radiation in the visible and infrared wavebands. **LANDSAT**, **SPOT** and **AVHRR** are examples of satellite platforms of this type. By sensing in the visible and infrared part of the electromagnetic spectrum the cloud brightness (visible) and temperature (thermal infrared) can be detected. This so-called 'brightness temperature' can then be related to rainfall intensity via calibration with point rainfall measurements, in a similar fashion to ground-based radar. One of the problems with this approach is that it is sometimes difficult to distinguish between snow reflecting light on the ground and clouds reflecting light in the atmosphere. They have similar brightness temperature values but need to be differentiated so that accurate rainfall assessment can be made.

Another form of satellite sensor that can be used is passive microwave. The earth emits microwaves (at a low level) that can be detected from space. When there is liquid water between the earth's surface and the satellite sensor (i.e. a cloud in the atmosphere) some of the microwaves are absorbed by the water. A satellite sensor can therefore detect the presence of clouds (or other bodies of water on the surface) as a lack of microwaves reaching the sensor. An example of a study using a satellite platform that can detect passive microwaves (SSM/I) is in Todd and Bailey (1995), who used the method to assess rainfall over the United Kingdom. Although there was some success in the method it is at a scale of little use to catchment scale hydrology as the best resolution available is around 10 × 10 km grid sizes.

Satellite remote sensing provides an indirect estimate of precipitation over an area but is still a long way from operational use. Studies have shown that it is an effective tool where there is poor rain gauge coverage (e.g. Kidd et al., 1998), but in countries with high rain gauge density it does not improve estimation of areal precipitation. What is encouraging about the technique is that nearly all the world is covered by satellite imagery so that it can be used in sparsely gauged areas. The new generation of satellite platforms being launched in the early twenty-first century will have multiple sensors on them so it is feasible that they will be measuring visible, infrared and microwave wavebands simultaneously. This will improve the accuracy considerably but it must be borne in mind that it is an indirect measure of precipitation and will still require calibration to a rain gauge set.

SUMMARY

Precipitation is the main input of water within a catchment water balance. Its measurement is fraught with difficulties and any small errors will be magnified enormously at the catchment scale. It is also highly variable in time and space. Despite these difficulties it is one of the most regularly measured hydrological variables, and good rainfall records are available for many regions in the world. Analysis of rainfall can be carried out with respect to trying to find a spatial average or looking at the intensity and duration of storm events. Although there are techniques available for estimating precipitation their accuracy is not such that it is superior to a good network of precipitation gauges.

ESSAY QUESTIONS

1   Describe the different factors affecting the spatial distribution of precipitation at differing scales.

2   How are errors in the measurement of rainfall and snowfall minimised?

3   Compare and contrast different techniques for obtaining an areal precipitation value (including surrogate measures).

4   Why is scale such an important issue in the analysis of precipitation in hydrology?

FURTHER READING

Bailey, T.C. and Gatrell, A.C. (1995) *Interactive spatial data analysis*. Longman, Harlow.
Gives a modern view of spatial analysis, not necessarily just for precipitation.

Barry, R.G. and Chorley, R. (1998) *Atmosphere, weather and climate* (7th edition). Routledge, London.
Explains the principles of precipitation generation well.

Cotton, W.R. and Anthes, R.A. (1989) *Storm and cloud dynamics*. Academic Press, San Diego.
More detail on processes of precipitation formation.

Sumner, G.N. (1988) *Precipitation: process and analysis*. J. Wiley & Sons, Chichester.
More detail on processes of precipitation formation and analysis from a meteorological point of view.

3

EVAPORATION

LEARNING OBJECTIVES

When you have finished reading this chapter you should have:

- An understanding of the process of evaporation and what controls its rate.
- A knowledge of the techniques for measuring evaporation directly.
- A knowledge of techniques used to estimate evaporation.

Evaporation is the transferral of liquid water into a gaseous state and its diffusion into the atmosphere. In order for this to occur there must be liquid water present and available energy from the sun or atmosphere. The importance of evaporation within the hydrological cycle depends very much on the amount of water present and the available energy, two factors determined by a region's climate. During the winter months evaporation is a minor component of the hydrological cycle in a humid temperate climate as there is very little available energy to drive the evaporative process. This changes completely during the summer when there is abundant available energy and evaporation becomes a major process. The converse of this can be seen in extremely hot, arid climates. Here there is often plenty of available energy to drive evaporation but very little water to

be evaporated. As a consequence the actual amount of evaporation is small.

It is the presence or lack of water at the surface that provides the major semantic distinction in definitions of the evaporative process. **Open water evaporation** (often denoted as E_o) is the evaporation that occurs above a body of water such as a lake, stream or the oceans. Figure 1.5 shows that this is the largest source of evaporation, in particular from the oceans. **Potential evaporation** (*PE*) is that which occurs over the land's surface, or would occur if the water supply is unrestricted. This occurs when a soil is wet and what evaporation is able to happen occurs without a lack of water supply. **Actual evaporation** (E_t) is that which occurs irrespective of the amount of available water. When conditions are very wet (e.g. during a rainfall event) E_t will equal *PE*, otherwise it will be less than *PE*. In hydrology we are most interested in E_o and E_t but normally require *PE* to calculate the E_t value.

All of these definitions have been concerned with 'evaporation over a surface'. In hydrology the surface is either water (river, lake, ponds, etc.) or the land. The evaporation above a land surface occurs in two ways – either as actual evaporation from the soil matrix or **transpiration** from plants. The combination of these two is often referred to as **evapotranspiration**, although the term *actual evaporation* is essentially the same (hence the *t* subscript in E_t). Transpiration from a plant occurs as part of photosynthesis and respiration. The rate of transpiration is controlled by the opening or closing of stomata in the leaf. Transpiration can be ascertained at the individual plant level by instruments measuring the flow of water up the trunk or stem of a plant. Different species of plants transpire at different rates but the fundamental controls are the available water in the soil and the plant's ability to transfer water from the soil to its leaves.

Traditionally evaporation is seen as the only loss within the water balance equation. This is erroneous for two reasons. The water balance equation is a mathematical description of the hydrological cycle and by definition there are no losses and gains within this cycle. What is meant by 'loss' is that evaporation is lost from the earth's surface, which is where hydrologists are mostly concerned with the water being. To a meteorologist, concerned with the atmosphere, evaporation can be seen as a gain. The second error with considering evaporation as a loss is that evaporation càn be reversed through condensation and **dewfall**. In this case the dewfall (or negative evaporation) is a gain to the earth's surface.

EVAPORATION AS A PROCESS

It has already been said that evaporation requires an energy source to transform liquid water into water vapour and an available water supply. There is one more precondition: that the atmosphere be dry enough to receive any water vapour produced. These are the three fundamental parts to an understanding of the evaporation process. This was first understood by Dalton (1766–1844), an English physicist who linked wind speed and the dryness of the air to the evaporation rate.

Available energy

The main source of energy for evaporation is from the sun. This is not necessarily in the form of direct radiation, it is often absorbed by a surface and then re-radiated at a different wavelength. The normal term used to describe the amount of energy received at a surface is **net radiation** ($Q*$), measured using a net radiometer. Net radiation is a sum of all the different heat fluxes found at a surface and can be described by the following equation:

$$Q* = Q_S \pm Q_L \pm Q_G$$

where Q_S is the sensible heat flux; Q_L is the latent heat flux and Q_G is the soil heat flux.

Sensible heat is that which can be sensed by instruments. This is easiest understood as the heat we feel as warmth. The sensible heat flux is the rate of flow of that sensible heat.

Latent heat is the energy required to produce a phase change from ice to liquid water, or liquid water to water vapour. When water moves from liquid to gas this is a negative flux (i.e. energy is lost) whereas the opposite phase change (gas to liquid) produces a positive heat flux.

The **soil heat flux** is heat released from the soil having been previously stored within the soil. This is frequently ignored as it tends to zero over a 24-hour period and is a relatively minor contributor to net radiation.

The energy available for evaporation, or the energy used in evaporation (latent heat) is balanced by energy from several sources: solar radiation, long-wave radiation, sensible heat and soil heat. Incoming solar radiation is filtered by the atmosphere so that not all the wavelengths of the electromagnetic spectrum are received at the earth's surface. Incoming radiation that reaches the surface is often referred to as short-wave radiation: visible light plus some bands of the infrared. This is not strictly true as clouds and water vapour in the atmosphere, plus trees and tall buildings above the surface, emit longer-wave radiation which also reaches the surface.

Outgoing radiation can be either reflected short-wave radiation or energy radiated back by the earth's surface. In the latter case this is normally in the infrared band and longer wavelengths and is referred to as long-wave radiation. This is a major source of energy for evaporation.

There are two other forms of available energy that under certain circumstances may be important sources in the evaporation process. The first is heat stored in buildings from an anthropogenic source (e.g. domestic heating). This energy source is often fuelled from organic sources and may be a significant addition to the heat budget in an urban environment, particularly in the winter months. The second additional source is **advective energy**. This is energy that originates from elsewhere (another region that may be hundreds or thousands of kilometres away) and has been transported to the evaporative surface (frequently in the form of latent heat) where it becomes available energy in the form of sensible heat. The best example of this is latent energy that arrives in cyclonic storm systems. In Chapter 1 it was explained that evaporating and condensing water is a major means of redistributing energy around the globe. The evaporation of water that contributes to cyclonic storms normally takes place over an ocean, whereas the condensation may occur a considerable distance away. At the time of evaporation, thermal energy (i.e. sensible heat) is transferred into latent energy that is then carried by the water vapour to the place of condensation where it is released as thermal energy once more. This 're-release' is often referred to as advective energy and may be a large energy source to drive further evaporation.

Water supply

Available water supply can be from water directly on the surface in a lake, river or pond. In this case it is open water evaporation (E_o). When the water is lying within soil the water supply becomes more complex. Soil water may evaporate directly, although it is normally only from the near surface. As the water is removed from the surface it sets up a moisture gradient that will draw water from deeper in the soil towards the surface, but it must overcome the force of gravity and the withholding force exerted by soil capillaries (see Chapter 5). In addition to this the water may be brought to the surface by plants using osmosis in their rooting system. The way that soil moisture controls the transformation from potential evaporation to actual evaporation is complex and will be discussed further later in this chapter.

The receiving atmosphere

Once the available water has been transformed into water vapour, using whatever energy source is available, it then must be absorbed into the atmosphere surrounding the surface. This process of *diffusion* requires that the atmosphere is not already saturated with water vapour and that there is enough buoyancy to move the water vapour away from the surface. These two elements can be assessed in terms of the **vapour pressure deficit** and atmospheric mixing.

Boyle's law tells us that the total amount of water vapour that may be held by a parcel of air is temperature and pressure dependent. The corollary of this is that for a certain temperature and air pressure it is possible to specify the maximum amount of water vapour that may be held by the parcel of air. We use this relationship to describe the **relative humidity** of the atmosphere (i.e. how close to fully saturated the atmosphere is). Another method of looking at the amount of water vapour in a parcel of air is to describe the *vapour pressure* and hence the *saturation vapour pressure* (see p. 14). The difference between the actual vapour pressure and the saturation vapour pressure is the *vapour pressure deficit* (vpd). The vpd is a measure of how much extra water vapour the atmosphere could hold assuming a constant temperature and pressure. The higher the vpd the more water can be absorbed from an evaporative surface.

Atmospheric mixing is a general term meaning how well a parcel of air is able to diffuse into the atmosphere surrounding it. The best indicator of atmospheric mixing is the wind speed at different heights above an evaporating surface. If the wind speed is zero the parcel of air will not move away from the evaporative surface and will 'fill' with water vapour. As the wind speed increases the parcel of air will be moved quickly on to be replaced by another, possibly drier, parcel ready to absorb more water vapour. If the evaporative surface is large (e.g. a lake) it is important that the parcel of air moves up into the atmosphere, rather than directly along at the same level, so that there is drier air replacing it. This occurs through turbulent diffusion of the air. There is a greater turbulence associated with air passing over a rough surface than a smooth one, something that will be returned to in the discussion of evaporation estimation.

One way of thinking about evaporation is in terms of a washing line. The best conditions for drying your washing outside are on a warm, dry, windy day. Under these circumstances the evaporation from your washing (the available water) is high due to the available energy being high (it is a warm day), and the receiving atmosphere mixes well (it is windy) and is able to absorb much water vapour. It is quite possible to have a warm and still day, or a warm and humid day when washing does not dry as well (i.e. the evaporation rate is low). Understanding evaporation in these terms allows us to think about what the evaporation rate might be for particular atmospheric conditions.

Evaporation above a vegetation canopy

Where there is a vegetation canopy the evaporation above this surface will be a mixture of transpiration and direct evaporation. Where the vegetation cover is dense, transpiration will be far more important than evaporation direct from the soil. This does not mean that straight evaporation will not be important though, particularly where there is water sitting on the vegetation during or after a rainfall event. This is normally referred to as *canopy interception* and will be discussed in more detail in Chapter 4.

MEASUREMENT OF EVAPORATION

In the previous chapter there has been much emphasis on the difficulties of measuring precipitation due to its inherent variability. All of these difficulties also apply to the measurement of evaporation, but they pale into insignificance when you consider that now we are dealing with measuring the rate at which a gas (water vapour) moves away from a surface. Concentrations of gases in the atmosphere are difficult to measure, and certainly there is no gauge that we can use to measure total amounts as we can for precipitation.

In each of the process chapters in this book there is an attempt to distinguish between measurement and estimation techniques. In the case of evaporation this distinction becomes extremely blurred. In reality almost all of the techniques used to find an evaporation rate are estimates, but some are closer to true measurement than others. In this section each technique will include a section on how close to 'true measurement' it is.

Direct micro-meteorological measurement

There are three main methods used to measure evaporation directly: the eddy fluctuation (or correlation), aerodynamic profile, and **Bowen ratio** methods. These are all micro-meteorological measurement techniques and details on them can be found elsewhere (e.g. Oke, 1987). An important point to remember about them all is that they are attempting to measure how much water is being evaporated above a surface, a very difficult task.

The eddy fluctuation method measures the water vapour above a surface in conjunction with a vertical wind speed and temperature profiles. These have to be measured at extremely short timescales (e.g. microseconds) to account for eddies in vertical wind motion. Consequently, extremely detailed micro-meteorological instrumentation is required with all instruments having a rapid response time. In recent years this has become possible with hot wire **anemometers** and extremely fine thermistor heads for thermometers. One difficulty is that you are necessarily measuring over a very small surface area and it may be difficult to scale up to something of interest to catchment-scale hydrology.

The aerodynamic profile (or turbulent transfer) method is based on a detailed knowledge of the energy balance over a surface. The fundamental idea is that by calculating the amount of energy available for evaporation the actual evaporation rate can be determined. The measurements required are changes in temperature and humidity giving vertical humidity gradients. To use this method it must be assumed that the atmosphere is neutral and stable, two conditions that are not always applicable.

The Bowen ratio method is similar to the aero-dynamic profile method but does not assume as much about the atmospheric conditions. The Bowen ratio is the ratio of sensible heat to latent heat and requires detailed measurement of net radiation, soil heat flux, temperature and humidity gradient above a surface. These measurements need to be averaged over a 30-minute period to allow the inherent assumptions to apply.

All of these micro-meteorological approaches to measuring evaporation use sophisticated instruments that are difficult to leave in the open for long periods of time. In addition to this they are restricted in their spatial scope (i.e. they only measure over a small area). With these difficulties it is not surprising that they tend to be used at the very small scale, mostly to calibrate estimation techniques (see pp. 37–40). They are reasonably accurate in the assessment of an evaporation rate, hence their use as a standard for the calibration of estimation techniques. The real problem for hydrology is that it is not a robust method that can be relied on for long periods of time.

Indirect measurement (water balance techniques)

Evaporation pans

The most common method for the measurement of evaporation is using an **evaporation pan** (see Figure 3.1). This is a large pan of water with a measuring stick or weighing device underneath that allows you to record how much water is lost through evaporation over a time period. This technique is actually a manipulation of the water balance equation, hence the terminology of a water balance technique. The evaporation pan has impervious sides and is not allowed to overflow, so no runoff (Q) term is required. It can also be assumed that any change in storage can be related to either evaporative loss or precipitation (i.e. there is no seepage or leakage out of the container). This means that the water balance equation can be rearranged as shown below:

$$E = \Delta S - P$$

If there is a precipitation gauge immediately adjacent to the evaporation pan then the P term can be accounted for, leaving only the change in storage (ΔS) to be measured as either a weight loss or a drop in water depth.

An evaporation pan is filled with water, hence you are measuring E_o, the open water evaporation. Although this is useful, there are severe problems with using this value as an indicator of actual evaporation (E_t) in a catchment. The first problem is that E_o will normally be considerably higher than

Evaporation pan

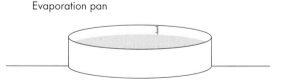

Figure 3.1 An evaporation pan. This sits above the surface (to lessen rain splash) and has a measurement rod to record loss of water.

E_t because the majority of evaporation in a catchment will be occurring over a land surface where the available water is contained within soil and may be limited. This will lead to a large overestimation of the actual evaporation. This factor is well known and consequently evaporation pans are rarely used in catchment water balance studies, although they are useful for estimating water losses from lakes and reservoirs.

There are problems with evaporation pans that make them problematic even for open water evaporation estimates. The size of the pan (up to 2 m diameter) makes them prone to the 'edge effect'. As warm air blows across a body of water it absorbs any water vapour evaporated from the surface. Numerous studies have shown that the evaporation rate is far higher near the edge of the water than towards the centre where the air is able to absorb less water vapour (this also applies to land surfaces). The small size of an evaporation pan means that the whole pan is effectively an 'edge' and will have a higher evaporation rate than a much larger body of water. A second, smaller, problem is that sides of the pan and the water inside will absorb radiation and warm up quicker than in a much larger lake, providing an extra energy source and greater evaporation rate. To overcome the edge effect, empirical (i.e. derived from measurement) coefficients can be used which link the evaporation pan estimates to larger water body estimates. Doorenbos and Pruitt (1975) give estimates for these coefficients that require extra information on wind speed and relative humidity (Goudie *et al.*, 1994).

Lysimeters

A **lysimeter** takes the same approach to measurement as the evaporation pan, the fundamental difference between them being that a lysimeter is filled with soil and vegetation as opposed to water (see Figure 3.2). This difference is important, as E_t rather than E_o is being measured. A lysimeter can also be made to blend in with the surrounding land cover, lessening the edge effect described for an evaporation pan.

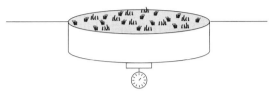

Weighing lysimeter

Figure 3.2 A weighing lysimeter sitting flush with the surface. The cylinder is filled with soil and vegetation similar to the surroundings.

There are many versions of lysimeters in use, but all use some variation of the water balance equation to estimate what the evaporation loss has been. One major difference from an evaporation pan is that a lysimeter allows percolation through the bottom, although the amount is measured. Percolation is necessary so that the lysimeter mimics as closely as possible the soil surrounding it; without any it would fill up with water. In the same manner as an evaporation pan it is necessary to measure the precipitation input immediately adjacent to the lysimeter. Assuming that the only runoff (Q) is through percolation the water balance equation for a lysimeter becomes:

$$E = \Delta S - P - Q$$

A lysimeter faces similar problems to a rain gauge in that it is attempting to measure the evaporation that would be lost from a surface if the lysimeter was not there. The difference from a rain gauge is that what is contained in the lysimeter should closely match the surrounding plants and soil. Although it is never possible to recreate the soil and plants within a lysimeter perfectly, a close approximation can be made and this represents the best efforts possible to measure evaporation. Lysimeters suffer from the same edge effect as evaporation pans, making it difficult to scale up from the single measurement to a catchment.

A *weighing lysimeter* has a weighing device underneath that allows any change in storage to be monitored. This can be an extremely sophisticated device (e.g. Campbell and Murray, 1990; Yang *et al.*,

---

Case study

A LYSIMETER USED TO MEASURE EVAPORATION FROM TUSSOCK

A narrow-leafed tussock grass (*Chinochloa rigida*, commonly called 'snow' or 'tall tussock') covers large areas of the South Island of New Zealand. A field study of a catchment dominated by snow tussock (Pearce *et al.*, 1984) has shown high levels of baseflow (i.e. high levels of **streamflow** between storm events). Mark *et al.* (1980), using a percolation gauge under a single tussock plant and estimating evaporation, have shown that the water balance can show a surplus. They suggested that this may be due to the tussock intercepting fog droplets that are not recorded as rainfall in a standard rain gauge (see Plate 3). The nature of a tussock leaf (long and narrow with a sharp point), would seem to be conducive to fog interception in the same manner as conifers intercepting fog. Another interpretation of the Mark *et al.* (1980) study is that the estimation of evaporation was incorrect. An understanding of the mechanisms leading to high **baseflow** levels is important for a greater understanding of hydrological processes leading to streamflow.

In order to investigate this further a large lysimeter was set up in two different locations. The lysimeter was 2 m in diameter and contained nine mature snow tussock plants in an undisturbed monolith, weighing approximately 8,000 kg. Percolating runoff was measured with a tipping-bucket mechanism and the whole lysimeter was on a beam balance giving a sensitivity of 0.054 mm. The rainfall was measured immediately adjacent to the lysimeter. Campbell and Murray (1990) show that although there were times when fog interception appeared to occur (i.e. the catch in the lysimeter was greater than that in the nearby rain gauge) this only accounted for 1 per cent of the total precipitation. The detailed measurements showed that the tussock **stomatal** or **canopy resistance** term was very high and that the plants had an ability to stop transpiring when the water stress became too high. The conclusion from the study was that snow tussocks are conservative in their use of water, which would appear to account for the high baseflow levels from tussock-covered catchments.

2000), where percolation is measured continuously using the same mechanism for a tipping-bucket rain gauge, weight changes are recorded continuously using a hydraulic pressure gauge, and precipitation is measured simultaneously. A variation on this is to have a series of small weighing lysimeters (such as small buckets) that can be removed and weighed individually every day to provide a record of weight loss. At the same time as weighing, the amount of percolation needs to be recorded. This is a very cheap way of estimating evaporation loss for an area using low technology.

Without any instrument to weigh the lysimeter (this is sometimes referred to as a *percolation gauge*) it must be assumed that the change in soil moisture over a period is zero and therefore evaporation equals rainfall minus runoff. This may be a reasonable assumption over a long time period such as a year where the soil storage will be approximately the same between two winters. An example of this type of lysimeter was the work of Law who investigated the effect that trees had on the water balance at Stocks Reservoir in Lancashire, UK (Law, 1956; see Case Study on p. 46).

A well-planned and executed lysimeter study probably provides the best information on evaporation that a hydrologist could find. However, it must be remembered that it is not evaporation that is being measured in a lysimeter – it is almost everything else in the water balance equation, with an assumption being made that whatever is left must be caused by evaporation. One result of this is that any errors in measurement of precipitation and/or percolation will transfer and possibly magnify into errors of evaporation measurement.

ESTIMATION OF EVAPORATION

The difficulties in measuring evaporation using either micro-meteorological instruments (problematic task used over short time periods and at the small scale) or water balance techniques (accumulated errors and small scale) has led to much effort being placed on estimating evaporation rather than trying to actually measure it. Some of the techniques outlined below are complicated and this sometimes leads hydrologists to believe that they are measuring, rather than estimating, evaporation. What they are actually doing is taking climatological variables that are known to influence evaporation and estimating evaporation rates from these: an estimation technique. The majority of research effort in this field has been to produce models to estimate evaporation; however, more recently satellite remote sensing has provided another method of estimating the evaporation flux.

The techniques described here represent a range of sophistication and they are certainly not all universally applicable. Almost all of these are concerned with estimating the potential evaporation over a land surface. As with most estimation techniques the hydrologist is required to choose the best techniques for the study situation. In order to help in this decision the various advantages and shortcomings of each technique are discussed.

Thornthwaite

Thornthwaite was an American engineer working in the 1940s and early 1950s. He derived an empirical model (i.e. derived from measurement not theoretical understanding) linking average air temperature to potential evaporation. This is an inherently sensible link in that we know air temperature is closely linked to both available energy and the ability of air to absorb water vapour.

The first part of the Thornthwaite estimation technique (Thornthwaite, 1944, 1954) derives a monthly heat index (i) for a region based on the average temperature t (°C) for a month.

$$i = \left(\frac{t}{5}\right)^{1.514}$$

These terms are then summed to provide an annual heat index I.

$$I = \sum_{j=1}^{12} i$$

Thornthwaite then derived an equation to provide evaporation estimates based on a series of observed evaporation measurements.

$$PE = 16b\left(\frac{10t}{I}\right)^{a}$$

The a and b terms in this equation can be derived in the following ways. Term b is a correction factor to account for unequal day length between months. Its value can be found by looking up tables based on the latitude of your study site. Term a is calibrated as a cubic function from the I term such as is shown below:

$$a = 6.7 \times 10^{-7} I^3 - 7.7 \times 10^{-5} I^2 + 0.018\,I + 0.49$$

The Thornthwaite technique is extremely useful as potential evaporation can be derived from knowledge of average temperature (often readily available from nearby weather stations) and latitude. There are drawbacks to its usage however, most notably that it only provides estimates of monthly evaporation. For anything at a smaller time-scale it is necessary to use another technique such as Penman (see pp. 38–40). There are also problems with using Thornthwaite's model in areas of high potential evaporation. The empirical nature of the model means that it has been calibrated for a certain set of conditions and that it may not be applicable outside these. The Thornthwaite model has been shown to underestimate potential evaporation in arid and semi-arid regions. If the model is being applied in conditions different to Thornthwaite's original calibration (humid temperate regions) it is advisable to find out if any researcher has published different calibration curves for the climate in question.

Penman

Penman was a British physicist working in the 1940s, 1950s and 1960s who derived a theoretical model of evaporation. Penman's first theoretical model was for open water evaporation and is shown below (Penman, 1948):

$$E_o = \frac{\frac{\Delta}{\gamma} Q^* + E_a}{\frac{\Delta}{\gamma} + 1}$$

where an empirical relationship states that:

$$E_a = 0.35\delta_e \left(0.5 + \frac{u}{100} \right)$$

and Q^* = net radiation (in units of mm/day)
  Δ = rate of increase of the saturation vapour deficit with temperature (see Figure 3.3)
  δ_e = vapour pressure deficit of the air
  γ = psychometric constant
  u = wind speed at 2 m

This formula requires observations of temperature, wind speed, vapour pressure (which can be derived from relative humidity) and net radiation and gives the evaporation in units of mm per day. These can be obtained from meteorological measurement (see p. 39). It is normal to use daily averages for these variables, although Shuttleworth (1988) has suggested that it should not be used for time steps of less than ten days. There are several different ways of presenting this formula, which makes it difficult to interpret between texts. The main difference is in whether the evaporation is a flux or an absolute rate.

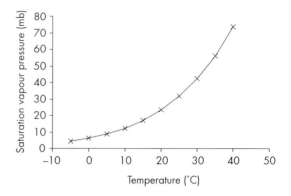

Figure 3.3 The relationship between temperature and saturation vapour pressure. This is needed to calculate the rate of increase of saturation vapour pressure with temperature (Δ).

In the equation above terms like 'net radiation' have been divided by the amount of energy required to evaporate 1 mm of water (density of water (ρ) multiplied by the latent heat of vaporisation (λ)) to turn them into water equivalents. This means the equation derives an absolute value for evaporation rather than a flux.

Penman continued his work to consider the evaporation occurring over a vegetated surface (Penman and Scholfield, 1951), while others refined the work (e.g. van Bavel, 1966). Part of this refinement was to include a term for aerodynamic resistance (r_a) to replace E_a. **Aerodynamic resistance** is a term to account for the way that the atmosphere mixes with evaporating air above it through turbulent mixing of the atmosphere. The rougher the canopy surface the greater degree of turbulent mixing will occur as air passing over the surface is buffeted around by protruding objects. As it is a resistance term, the higher the value the greater the resistance to mixing; therefore a forest actually has a lower value of r_a than smoother pasture. Taking these alterations into account, and presenting the results as a water flux, the Penman equation can be written as shown:

$$PE = \frac{Q^* \Delta + \rho c_p \delta_e / r_a}{\lambda(\Delta + \gamma)}$$

where

$$r_a = \frac{1}{\kappa^2 u}\left(\ln\left(\frac{z-d}{z_0}\right)\right)^2$$

and $Q*$ = net radiation (W/m^2)
Δ = rate of increase of the saturation vapour pressure with temperature (mb/°C) (see Figure 3.3)
ρ = density of air (kg/m^3)
c_p = specific heat of air at constant pressure ($\approx$ 1,005 J/kg)
δ_e = vapour pressure deficit of the air (mb)
λ = latent heat of vaporisation of water (J/kg) (see Figure 3.4)
γ = psychometric constant ($\approx$ 0.66)
r_a = aerodynamic resistance to transport of water vapour (s/m)
κ = Von Karman constant ($\approx$ 0.41)
u = wind speed above canopy (m/s)
z = height of anemometer (m)
d = **zero plane displacement** (the height within a canopy at which wind speed drops to zero) (m)
z_0 = roughness height (height of vegetation) (m)

Although this formula looks complicated it is actually rather simple. It is possible to split the equation into two separate parts that conform to the understanding of evaporation already discussed. The available energy term is predominantly assessed through the net radiation ($Q*$) term. Other terms in the equation relate to the ability of the atmosphere to absorb the water vapour (Δ, ρ, c_p, δ_e, λ, γ) and the rate at which diffusion will absorb the water vapour into the atmosphere (κ, u, z_0 etc).

When using the Penman equation there are only four variables requiring measurement: net radiation, wind speed above the canopy, atmospheric humidity and temperature, which when combined will provide vapour pressure deficit (see Figures 3.3–3.5). Every other term in the equation is either a constant, a simple relationship from another variable or can be measured once. Of these four

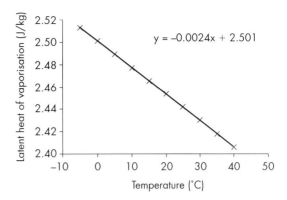

Figure 3.4 The relationship between temperature and latent heat of vaporisation.

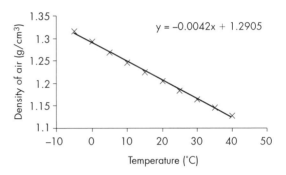

Figure 3.5 The relationship between air temperature and the density of air.

variables net radiation is the hardest to obtain from meteorological stations as net radiometers are not common. There are methods of estimating net radiation from measurements of incoming solar radiation, surface **albedo** (or reflectivity) and day length (see Oke, 1987).

The modified Penman equation provides estimates of potential evaporation at a surface for time intervals much less than the monthly value from Thornthwaite. This makes it extremely useful to hydrology and it is probably the most widely used method for estimating potential evaporation values.

There are problems with the Penman equation which make it less than perfect as an estimation technique. The assumption is made that the soil heat flux is unimportant in the evaporation energy

budget. This is often the case but is an acknowl-
edged simplification that may lead to some overall
error, especially when the time step is less than one
day. It is normal practice to use Penman estimates
at the daily time step; however, in some modelling
studies they are used at hourly time steps.

One major problem with the Penman equation
relates to its applicability in a range of situations
and in particular in the role of advection, as discussed on p. 32. This is where there are other energy
sources available for evaporation that cannot be
assessed from net radiation. Calder (1990) shows the
results from different studies in the UK uplands
where evaporation rates vastly exceed the estimates
provided by the Penman equation. The cause of this
discrepancy is the extra energy provided by cyclonic
storms coming onto Britain from the Atlantic
Ocean, something that is poorly accounted for in
the Penman equation. The sensible heat transfer
function within the Penman equation does account
for some advection but not if it is a major energy
source driving evaporation. This does not render the
Penman approach invalid; rather, in applying it
the user must be sure that net radiation is the main
source of energy available for evaporation.

Simplifications to Penman

There have been several attempts made to simplify
the Penman equation for widespread use. Slatyer
and McIlroy (1961) separated out the evaporation
caused by sensible heat and advection from that
caused by radiative energy. Priestly and Taylor
(1972) derived a simplified Penman formula for
use in the large-scale estimation of evaporation, in
the order of 'several hundred kilometres' where it
can be argued that large-scale advection is not
important. Their formula for potential evaporation
is shown below:

$$PE = \alpha \frac{\left(Q^* - Q_G\right)\Delta}{\lambda\left(\Delta + \gamma\right)}$$

where Q_G is the soil heat flux term (ignored by
Penman) and α is the Priestly–Taylor parameter, all

other parameters being as defined earlier. The α term
is an approximation of the sensible heat transfer and
was estimated by Priestly and Taylor (1972) to have
a value of 1.26 for saturated land surfaces, oceans and
lakes – that is to say, that the sensible heat transfer
accounts for 26 per cent of the evaporation over and
above that from available energy. This value of α has
been shown to vary away from 1.26 (e.g. α = 1.21
in Clothier et al., 1982) but to generally hold true
for large-scale areas without a water deficit.

Penman–Monteith

Monteith (1965) derived a further term for the
Penman equation so that actual evaporation from a
vegetated surface could be estimated. His work
involved adding a canopy resistance term (r_c) into
the Penman equation so that it takes the following
form

$$E_t = \frac{Q^*\Delta + \rho c_p \delta_e / r_a}{\lambda\left(\Delta + \gamma\left(1 + \dfrac{r_c}{r_a}\right)\right)}$$

Looking at the Penman–Monteith equation you
can see that if r_c equals zero then it reverts to the
Penman equation (i.e. actual evaporation equals
potential evaporation). If the canopy resistance is
high the actual evaporation rate drops to less
than potential. Canopy resistance represents the
ability of a vegetation canopy to control the rate of
transpiration. This is achieved through the opening
and closing of stomata within a leaf, hence r_c is
sometimes referred to as stomatal resistance. Various
researchers have established canopy resistance values
for different vegetation types (e.g. Szeicz et al.,
1969), although they are known to vary seasonally
and in some cases diurnally. Rowntree (1991)
suggests that for grassland under non-limiting
moisture conditions the range of r_c should fall somewhere between 60 and 200 s/m. The large range is
a reflection of canopy resistance being influenced by
a plant's physiological response to variations in soil
moisture condition and climatological conditions.

Simple estimation of E_t from *PE* and soil moisture

The relationship between actual evaporation (E_t) and potential evaporation (*PE*) is driven by the availability of water. Over a land surface the availability of water can be estimated from the soil moisture content (see Chapter 5). At a simple level it is possible to estimate the relationship between potential and actual evaporation using soil moisture content as a measured variable (see Figure 3.6). In Figure 3.6 a value of 1 on the y-axis corresponds to actual precipitation equalling potential evaporation (i.e. available water is not a limiting factor on the evaporation rate). The exact position where this occurs will be dependent on the type of soil and plants on the land surface, hence the lack of units shown on the x-axis and the two different curves drawn. This type of simple relationship has been effective in determining actual evaporation rates in a crude model of soil water budgeting (e.g. Davie *et al.*, 2001) but cannot be relied on for accurate modelling studies. It provides a very crude estimate of actual evaporation from knowledge of soil moisture and potential evaporation.

Remote sensing of evaporation

Water vapour is a greenhouse gas and therefore it interferes with radiation (i.e. absorbs and reradiates) from the earth's surface. Because of this the amount of water vapour in the atmosphere can be estimated using satellite remote sensing, particularly using passive microwave sensors. The difficulty with using this information for hydrology is that it is at a very large scale (often continental) and is concerned with the whole atmosphere not the near surface. In order to utilise satellites for estimation of evaporation a combined modelling and remote sensing approach is required. Burke *et al.* (1997) describe a combined Soil–Vegetation–Atmosphere–Transfer (SVAT) model that is driven by remotely sensed data. This type of approach can be used to estimate evaporation rates over a large spatial area relatively easily. Mauser and Schädlich (1998) provide a review of evaporation modelling at different scales using remotely sensed data.

Mass balance estimation

In the same manner that evaporation pans and lysimeters estimate evaporation rates, evaporation at the large scale (catchment or lake) can be estimated through the water balance equation. This is a relatively crude method, but it can be extremely effective over a large spatial and/or long temporal scale. The method requires accurate measurement of precipitation and runoff for a catchment or lake. In the case of a lake, change in storage can be estimated through lake-level recording and knowledge of the surface area. For a catchment it is often reasonable to assume that change in storage is negligible over a long time period (e.g. one year) and therefore the evaporation is precipitation minus runoff.

SUMMARY

The evaporation process involves the transfer of water from a liquid state into a gaseous form in the atmosphere. For this to happen requires an available energy source, a water supply and the ability of the atmosphere to receive it. Evaporation is difficult to measure directly and there are various estimation techniques.

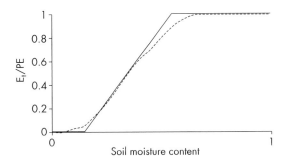

Figure 3.6 A hypothetical relationship between the measured soil moisture content and the ratio of actual evaporation to potential evaporation.

These range from water budget techniques, such as evaporation pans and lysimeters, to modelling techniques, such as the Penman–Monteith equation. As a process, evaporation suffers from the same problems with measurement and estimation as does precipitation (i.e. extreme variability in space and time). This variability leads to difficulties in moving from point measurements to areal estimates such as are required for a catchment study. These can be overcome by using spatial averaging techniques or using evaporation estimations that assume a large base area (e.g. Priestly–Taylor).

FURTHER READING

Brutsaert, W. (1982) *Evaporation into the atmosphere: theory, history, and applications*. Kluwer, Dordrecht.
A detailed overview of the evaporation process.

Calder, I.R. (1990) *Evaporation in the uplands*. J. Wiley & Sons, Chichester.
Although concerned primarily with upland evaporation it covers the issues of estimation well.

ESSAY QUESTIONS

1 Give a detailed account of the factors that influence the rate of evaporation above a certain point on the earth's surface.

2 Compare and contrast the use of evaporation pans and lysimeters for measuring evaporation.

3 Explain why are there more evaporation estimation techniques than there is evaporation measurement.

4 Outline the major evaporation estimation techniques and compare their effectiveness for your local environment.

4

INTERCEPTION

LEARNING OBJECTIVES

When you have finished reading this chapter you should have:

■   An understanding of the process of precipitation interception by a canopy.
■   A knowledge of the techniques for measuring the amount of canopy interception in a canopy.
■   A knowledge of techniques used to estimate canopy interception.

To a hydrologist who is fundamentally interested in the amount of water flowing down a river, vegetation canopy is a barrier for precipitation to cross before reaching the soil and possibly making its way to the river. To the meteorologist interested in earth–atmosphere interaction, vegetation creates a blurred surface for evaporation to occur from. With these two differing viewpoints it is perhaps not surprising that interception is sometimes not perceived as a separate process within the hydrological cycle, but rather an amalgam of precipitation and evaporation. Here canopy interception will be treated separately, although many of the fundamental concepts have already been introduced in Chapter 3. Of course it is not restricted to just vegetation, canopies are produced by buildings and other structures from which water can be evaporated. But

it is vegetation that has the largest influence on the hydrological cycle, purely because of the amount of land surface that is covered by it.

CANOPY RAINFALL PARTITIONING

Once rain falls onto a vegetation canopy it can be said to partition the water into separate modes of movement: **throughfall**, **stemflow** and interception loss. This is illustrated in Figure 4.1.

Throughfall

This is the water that falls to the ground either directly, through gaps in the canopy, or indirectly, having dripped off leaves, stems or branches.

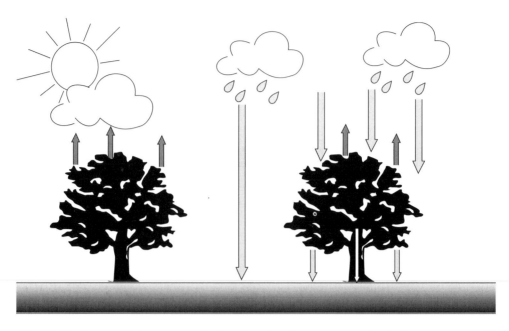

Figure 4.1 Rainfall above and below a canopy. Indicated on the diagram are stemflow (white arrow on trunk); direct and indirect throughfall (lightly hatched arrow), and interception loss (upward-facing darker arrow).

The amount of *direct throughfall* is controlled by the canopy coverage for an area, a measure of which is the leaf area index (LAI). LAI is actually the ratio of leaf area to ground surface area and consequently has a value greater than one when there is more than one layer of leaf above the ground. When the LAI is less than one you would expect some direct throughfall to occur. When you shelter under a tree during a rainstorm you are trying to avoid the rainfall and direct throughfall. The greater the surface area of leaves above you, the more likely it is that you will avoid getting wet from direct throughfall.

The amount of *indirect throughfall* is also controlled by the LAI, in addition to the **canopy storage capacity** and the rainfall characteristics. Canopy storage capacity is the volume of water that can be held by the canopy before water starts dripping as indirect throughfall. The canopy storage capacity is controlled by the size of trees, plus the area and water-holding capacity of individual leaves. Rainfall characteristics are an important control on indirect throughfall as they dictate how quickly the canopy

storage capacity is filled. Experience of standing under trees during a rainstorm should tell you that intensive rainfall quickly turns into indirect throughfall (i.e. you get wet!), whereas light showers frequently do not reach the ground surface at all. In reality canopy storage capacity is a rather nebulous concept. Canopy characteristics are constantly changing and it is rare for water on a canopy to fill up completely before creating indirect throughfall. This means that indirect throughfall occurs before the amount of rainfall equals the canopy storage capacity, making it difficult to gauge exactly what the storage capacity is.

Stemflow

Stemflow is the rainfall that is intercepted by stems and branches and flows down the tree trunk into the soil. Although measurements of stemflow show that it is a small part of the hydrological cycle (normally 2–10 per cent of above canopy rainfall; Lee, 1980) it can have a much more significant role. Stemflow acts like a funnel (see Figure 4.2), collecting water from a large area of canopy but delivering it to the soil in a much smaller area: the surface of the trunk at the base of a tree. This is most obvious for the deciduous oak-like tree illustrated in Figure 4.2, but it still applies for other structures (e.g. conifers) where the area of stemflow entry into the soil is far smaller than the canopy catchment area for rainfall. At the base of a tree it is possible for the water to rapidly enter the soil through flow along roots and other macropores surrounding the root structure. This can act as a rapid conduit of water sending a significant pulse into the **soil water**.

The complex nature of a forest topography means that trees act as scavengers of air pollutants. As rain falls onto the tree, salts that have formed on leaves and branches may be dissolved by the water, making the stemflow pollutant-rich. This has been observed in field studies, particularly near the edge of tree stands (Neal *et al.*, 1991).

Interception loss

While water sits on the canopy, prior to indirect throughfall or stemflow, it is available for evaporation, referred to as *interception loss*. It is the role of wet leaf evaporation (interception loss + transpiration) that makes afforested areas greater users of water than pasture land (see Case Study on p. 46). There are two types of factors influencing the amount of interception loss from a particular canopy: plant physiology and meteorological.

Plant physiology factors include the storage capacity (discussed on p. 44) and the drainage charac-

Figure 4.2 The funnelling effect of a tree canopy on stemflow.

teristics of the canopy. The morphology of leaf and bark on a tree are important factors in controlling how quickly water drains towards the soil. If leaves are pointed upwards then there tends to be a rapid drainage of water towards the stem. Sometimes this is a genetic strategy by a plant in order to harvest as much water as possible (e.g. rhubarb and gunnera plants). Large broadleaved plants, such as oak (*Quercus*) tend to hold water well on their leaves while needled plants can hold less per leaf (although they normally have more leaves). Seasonal changes make a large difference within deciduous forests, with far greater interception losses when the trees have leaves than without. Durocher (1990) found that trees with smoother bark such as beech (*Fagus*) had higher rates of stemflow as the smoothness of bark tends to enhance drainage towards stemflow. Table 4.1 illustrates the influence of plant morphology through the variation in interception found in different forest types and ages.

Table 4.1 Interception measurements in differing forest types and ages

Tree type	Age	Interception (mm)	Percentage of annual precipitation
Deciduous hardwoods	100	254	12
Pinus strobus (white pine)	10	305	15
Pinus strobus	35	381	19
Pinus strobus	60	533.4	26

Source: From Hewlett and Nutter (1969)

Meteorological factors affecting the amount of interception loss are either rainfall characteristics or influences on evaporation rates (see Chapter 3). The rate at which rainfall occurs (intensity), and storm duration, is critical in controlling the interception loss. The longer water stays on the canopy the greater the amount of interception loss. Also important will be the frequency of rainfall. Does the canopy have time to dry out between rain events? If so, then the interception amount is likely to be higher.

The amount of interception loss from an area is climate dependent. Calder and Newson (1979) use

---

Case study

TREES VS. GRASSLAND

If you stand watching a stand of trees during a warm summer shower it is common enough to see what appears to be clouds forming above the rain (see Plate 4). For many years it was believed that somehow trees attract rainfall and that cloud-forming was evidence of this phenomenon. This idea was taken further so that it became common practice to have forestry as a major land use in catchments that were being used to collect water for potable supply. In actual fact the cloud formation that is visible above a forest is a result of evaporation occurring from water sitting on the vegetation (intercepted rainfall). This 'wet leaf evaporation' can be perceived as a loss to the hydrologist as it does not reach the soil surface and contribute to possible streamflow. Throughout the latter half of the twentieth century there was considerable debate on how important wet leaf evaporation is.

One of the first pieces of field research to promote the idea of canopy interception being important was undertaken at Stocks Reservoir, Lancashire, UK. Law (1956) studied the water balance of an area covered with conifers (Sitka spruce) and compared this to a similar area covered with grassland. The water balance was evaluated for areas isolated by impermeable barriers with evaporation left as the residual (i.e. rainfall and runoff were measured and soil moisture assumed constant by looking at yearly values). Law found that the evaporation from the forested area was far greater than that for the pasture and he speculated that this was caused by wet leaf evaporation – in particular that the wet leaf evaporation was far greater from the forested area as there was a greater storage capacity for the intercepted water. Furthermore, Law went on to calculate the amount of water 'lost' to reservoirs through wet leaf vegetation and suggested a compensation payment from the forestry owners to water suppliers.

Conventional hydrological theory at the time suggested that wet leaf evaporation was not an important part of the hydrological cycle because it compensated for the reduction in transpiration that occurred at the same time (e.g. Leyton and Carlisle, 1959; Penman, 1963). In essence it was believed that the evapotranspiration rate stayed constant whether the canopy was wet or dry.

Following the work of Law, considerable research effort was directed towards discovering whether the wet leaf/dry leaf explanation was responsible for discrepancies in the water balance between grassland and forest catchments. Rutter (1967) and Stewart (1977) found that wet leaf evaporation in forests may be up to three or four times that from dry leaf. In contrast to this, other work has shown that on grassland wet leaf evaporation is approximately equal to dry leaf (McMillan and Burgy, 1960; McIlroy and Angus, 1964). In addition, transpiration rates for pasture have been found to be similar to that of forested area. When all this evidence is added up it confirms Law's work that forested areas 'lose' more rainfall through evaporation of intercepted water than grassland areas.

an amalgamation of different forest interception studies to show that there is a higher interception ratio (the interception loss divided by above-canopy rainfall) in drier than in wetter climates.

Interception gain

In some circumstances it is possible that there is an interception gain from vegetation. In the Bull Run catchment, Oregon, USA it has been shown that the water yield after timber harvesting was significantly less than prior to the trees being logged (Harr, 1982; Ingwersen, 1985). This is counter to the majority of catchment studies reported by Bosch and Hewlett (1982) which show an increase in water yield as forests are logged. The reason for the loss of water with the corresponding loss of trees in Oregon is to do with the particular circumstances of the catchment. Fog from the cold North Pacific, with no accompanying rain, is a common feature and it is believed that the trees intercept fog particles, creating 'fog drip' which is a significant part of the water balance. Fog droplets are extremely small and Ingwersen (1985) has suggested that the sharp ends of needles on pine trees act as condensation nuclei, promoting the growth of larger droplets that fall to the ground (see an example of fogdrip from tussock leaves in Plate 3). When the trees are removed there are no condensation nuclei (or far fewer) on the resultant vegetation so the water remains in the atmosphere and is 'lost' in terms of water yield. The overall result of this is that the removal of trees leads to less water in the river; this runs counter to the evidence provided in the Case Study for this chapter.

MEASURING CANOPY INTERCEPTION

The most common method of assessing the amount of canopy interception is to measure the precipitation above and below a canopy and assume that the difference is from interception. Stated in this way it sounds a relatively simple task but in reality it is fraught with difficulty and error. Durocher (1990) provides a good example of the instrumentation necessary to measure canopy interception, in this case for a deciduous woodland plot.

Above-canopy precipitation

To measure above-canopy precipitation a rain gauge may be placed on a tower above the canopy. The usual rain gauge errors (see Chapter 2) apply here, but especially the exposure to the wind. As described in Chapter 3, the top of a forest canopy tends to be rough and is very good for allowing turbulent transfer of evaporated water. The turbulent air is not so good for measuring rainfall! An additional problem for any long-term study is that the canopy is not static; the tower needs to be raised every year so that it remains above the canopy.

One way around the tower problem is to place a rain gauge in a nearby clearing and assume that what falls there is the same amount as directly above the canopy nearby. This is often perfectly reasonable to assume, particularly for long-term totals, but care must be taken to ensure the clearing is large enough to avoid obstruction from nearby trees (see Figure 2.7).

Throughfall

Throughfall is the hardest part of the forest hydrological cycle to measure. This is because a forest canopy is normally variable in density and therefore throughfall is spatially heterogeneous. One common method is to place numerous rain gauges on the forest floor in a random manner. If you are interested in a long-term study then it is reasonable to keep the throughfall gauges in fixed positions. However, if the study is investigating individual storm events then it is considered best practice to move the gauges to new random positions between storm events. In this way the throughfall catch should not be influenced by gauge position. To derive an average throughfall figure it is necessary to come up with a spatial average in the same manner as for areal rainfall estimates (Chapter 3).

Figure 4.3 Throughfall trough sitting beneath a pine tree canopy. This collects rain falling through the canopy over the area of the trough. It is sloping so that water flows to a collection point.

To overcome the difficulty of a small sampling area (rain gauge) measuring something notoriously variable (throughfall), some investigators have used either troughs or plastic sheeting. Troughs collect over a greater area and have proved to be very effective (see Figure 4.3). Plastic sheeting is the ultimate way of collecting throughfall over a large area, but has several inherent difficulties. The first is purely logistical in that it is difficult to install and maintain, particularly to make sure there are no rips. The second is that by having an impervious layer above the ground there is no, or very little, water entering the soil. This might not be a problem for a short-term study but is over the longer term, especially if the investigator is interested in the total water budget. It may also place the trees under stress through lack of water, thus leading to an altered canopy.

Stemflow

The normal method of measuring stemflow is to place collars around a tree trunk that capture all the water flowing down the trunk. On trees with smooth bark this may be relatively simple but is very difficult on rough bark such as found on many conifers. It is important that the collars are sealed to the tree so that no water can flow underneath and that they are large enough to hold all the water flowing down the trunk. The collars should be sloped to one side so that the water can be collected or measured in a tipping-bucket rain gauge. Maintenance of the collars is very important as they easily clog up or become appropriate resting places for forest fauna such as slugs!

ESTIMATING CANOPY INTERCEPTION

As with evaporation the main effort in estimating interception has been using numerical models. Regression models that link rainfall to interception loss based on a measured data set have been developed for many different types of vegetation canopy (see Zinke [1967] and Massman [1983] for examples and reviews of these types of model). Some of these models used logarithmic or exponential terms in the equations but they all rely on having regression coefficients based on the vegetation type and climatic regime.

A more detailed modelling approach is the Rutter model (Rutter *et al.*, 1971, 1975) which calculates an hourly water balance within a forest stand. The water balance is calculated taking into account the rate of throughfall, stemflow, interception loss through evaporation and canopy storage. In order to use the model a detailed knowledge of the canopy characteristics is required. In particular the canopy storage and drainage rates from throughfall are required to be known; the best method for deriving these is through empirical measurement. The Rutter model treats the canopy as a single large leaf, although it has been adapted to provide a three-dimensional canopy (e.g. Davie and Durocher, 1997) that can then be altered to allow for changes and growth in the canopy.

At present, remote sensing techniques are not able to provide reasonable estimates of canopy interception. They do provide some useful information that can be incorporated into canopy interception models but cannot provide the detailed difference between above- and below-canopy rainfall. In particular, satellites can give good information on the type of vegetation and its degree of cover. Particular care needs to be taken over the term 'leaf area index' when reading remote sensing literature. Analysis of remotely sensed images can provide a good indication of the percentage vegetation cover for an area, but this is not necessarily the same as leaf area index – although it is sometimes referred to as such. Leaf area index is the surface area of leaf cover above a defined area divided by the surface area defined. As there are frequently layers of vegetation above the ground, the leaf area index frequently has a value higher than one. The percentage vegetation cover cannot exceed one (as a unitary percentage) as it does not consider the third dimension (height).

SNOWFALL INTERCEPTION

In the same manner that rainfall may be intercepted by a canopy, so can snow. The difference between the two is in the mass of water held and the duration of storage (Lundberg and Halldin, 2001). The amount of intercepted snow is frequently much higher than for rainwater and it is held for much longer. This may be available for evaporation through sublimation (moving directly from a solid to a gas) or release later in snow melt (see Chapter 5). Hedstrom and Pomeroy (1998) point out that the mass of snow held by interception is controlled by the tree branching structure, leaf area and tree species. In countries such as Canada and Russia there are extensive forests in regions dominated by winter snowfall. Some studies have shown as much as 20–50 per cent of gross precipitation being intercepted and evaporated (Lundberg and Halldin, 2001). These figures indicate that a consideration of snowfall interception is critical in these regions. Lundberg and Halldin (2001) provide a recent review of measuring snow interception and modelling techniques.

SUMMARY

The forest canopy partitions rainfall into components that move at different rates towards the soil surface. The nature of the canopy (leaf size distribution and leaf area index) determines the impact that a canopy has on the water balance equation. In general, water is lost from the catchment through evaporation off wet leaves, but this is not always the case – there are cases where a tree canopy leads to more water in the catchment. The importance of canopy interception in a catchment water balance is dependent on the size and extent of vegetation cover found within a watershed.

ESSAY QUESTIONS

1   **Assess the relative importance of throughfall, stemflow and interception for two different canopy covers (e.g. deciduous and coniferous woodland).**

2   **Describe the field experiment (including equipment) required to measure the water balance beneath a forest canopy.**

3   **Discuss the role of spatial scale in assessing the importance of a forest canopy within a watershed.**

FURTHER READING

Calder, I.R. (1990) *Evaporation in the uplands*. J. Wiley & Sons, Chichester.
Although concerned primarily with upland evaporation it covers interception studies well.

Lee, R. (1980) *Forest hydrology*. Columbia University Press, New York.
An overview of forest hydrological processes with some detail on interception.

5

STORAGE

LEARNING OBJECTIVES

When you have finished reading this chapter you should have:

- An understanding of the role of water stored below the ground (in both the saturated and unsaturated zones).
- An understanding of the role of snow and ice acting as a store for water.
- A knowledge of the techniques for measuring snow and ice and water beneath the ground.
- A knowledge of techniques used to estimate the amount of water stored as soil moisture, groundwater and snow and ice.

The water balance equation, explained in Chapter 1, contains a storage term (S). Within the hydrological cycle there are several areas where water can be considered to be stored, most notably soil moisture, groundwater, snow and ice and, to a lesser extent, lakes and reservoirs. It is tempting to see stored water as static, but in reality there is considerable movement involved. The use of a storage term is explained in Figure 5.1 where it can be seen that there is an inflow, an outflow and a movement of water between the two. The inflow and outflow do not have to be equal over a time period; if not, then there has been a *change in storage* (ΔS). The critical point is that at all times there is some water stored, even if it is not the same water throughout a measurement period.

This definition of stored water is not perfect as it could include rivers as stored water in addition to groundwater etc. The distinction is often made on the basis of flow rates (i.e. how quickly the water moves while in storage). There is no critical limit to say when a deep, slow river becomes a lake, and likewise there is no definition of how slow the flow has to be before becoming stored water. It relies on an intuitive judgement that slow flow rates constitute stored water.

The importance of stored water is highlighted by the fact that it is by far the largest amount of fresh

Figure 5.1 Illustration of the storage term used in the water balance equation.

water in or around planet earth. The majority of this is either in snow and ice (particularly the polar ice caps) or groundwater. For many parts of the world groundwater is a major source of drinking water, so knowledge of amounts and replenishment rates is important for hydrological management. By definition, stored water is slow moving so it is particularly prone to contamination by pollutants. The three 'Ds' of water pollution control (dilution, dispersion and degradation; see Chapter 8) all occur at slow rates in stored water, making pollution management a particular problem. When this is combined with the use of these waters for potable supply, an under-

standing of the hydrological processes occurring in stored water is very important.

In this chapter two major stores of water are described: water beneath the earth's surface in the unsaturated and saturated zones; and snow and ice.

WATER BENEATH THE EARTH'S SURFACE

One way of considering water beneath the earth's surface is to divide it between the saturated and unsaturated zones (see Figure 5.2). Water in the saturated zone is referred to as *groundwater* and occurs beneath a **water table**. This is also referred to as water in the **phreatic zone**.

Water in the unsaturated zone is referred to as *soil water* and occurs above a water table. This is also referred to as water in the **vadose zone**.

In reality all water beneath the surface is groundwater, but it is convenient to distinguish between the saturated and unsaturated zones and maintain the terminology used by other hydrologists. As shown in Figure 5.2, there is movement of water

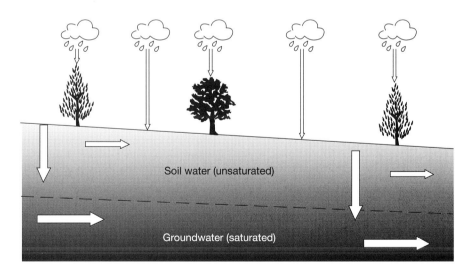

Figure 5.2 Water stored beneath the earth's surface. Rainfall infiltrates through the unsaturated zone towards the saturated zone. The dashed line represents the water table, although, as the diagram indicates, this is actually a gradual transition from unsaturated to fully saturated.

through both vertical infiltration and horizontal flow (in reality this is a combined vector effect). It is important to realise that this occurs in both the unsaturated and saturated zones, although at a slower rate in the unsaturated.

Water in the unsaturated zone

The majority of water in the unsaturated zone is held in soil. Soil is essentially a continuum of solid particles (minerals, organic matter), water and air. Consideration of water in the soil starts with the control over how much water enters a soil during a certain time interval: the **infiltration rate**. The rate at which water enters a soil is dependent on the current water content of the soil and the ability of a soil to transmit the water.

Soil water content

Soil water content is normally expressed as a volumetric water content or fraction and given the Greek symbol theta (θ). This is the volume of water in a soil sample (V_w) divided by the total volume of soil sample (V_t).

$$\theta = \frac{V_w}{V_t}$$

This is normally kept as unitary percentage (i.e. 1 = 100 per cent). As θ is a volume divided by a volume it has no units, although if is sometimes denoted as $m^3 m^{-3}$ (or m^3/m^3). Volumetric water content of a soil is effectively a depth ratio that is easily related to other equivalent depths such as rainfall and evaporation (when expressed in mm depth).

Soil water content is sometimes described by *gravimetric water content* (G). *Gravimetric soil moisture content* is a ratio of the weight of water in a soil to the overall weight of the soil:

$$G = \frac{M_w - M_d}{M_d}$$

As the density of water is very close to 1 g/cm^3 (temperature dependent; see Figure 1.3) the weight of water is often assumed to be the same as the volume of water. The same cannot be said for soil: the density depends on the mineralogy and packing of particles so that the volume does not equal the weight. Because of this, gravimetric soil moisture content is not the same as **volumetric soil moisture content**, and care must be taken in distinguishing between them as they are not interchangeable terms.

A third way of expressing soil water content is as a percentage of saturated. **Saturated water content** is the maximum amount of water that the soil can hold. It is sometimes referred to as the **porosity**, which assumes that the water fills all the pore space within a soil. Soil water content as percentage of saturated is a useful method of telling how wet the soil actually is.

Other terms used in the description of soil moisture content are **field capacity**, **soil moisture deficit** and **wilting point**. *Field capacity* is the actual maximum water content that a soil can hold under normal field conditions. This is often less than the porosity as the water does not fill all the pore space and is constantly under the influence of gravity. *Soil moisture deficit* is the amount of water required (in mm depth) to fill the soil up to field capacity. This is an important hydrological parameter as it is often assumed that all rainfall infiltrates into a soil until the moisture content reaches field capacity. The soil moisture deficit gives an indication of how much rain is required before saturation, and therefore when overland flow may occur (see Chapter 6). *Wilting point* is a term derived from agriculture and refers to the soil water content when plants start to die back (wilt). This is significant for hydrology as beyond this point the plants will no longer transpire.

Ability to transmit water

The ability of a soil to transmit water is dependent on the pore spaces within it and most importantly on the connections between pores. The measure of a soil's ability to transmit water is **hydraulic** conductivity. When the soil is wet, water flows through the soil at a rate controlled by the saturated hydraulic conductivity (K_{sat}). The rate of flow is best described by Darcy's law, which will be explained to a fuller extent later in this chapter in the section on groundwater. When the soil is unsaturated it transmits water at a much slower rate, which is described by the Richards approximation of Darcy's law. This links the flow rate to the soil water content in a logarithmic manner so that a small change in water content can lead to a rapid increase in flow rate.

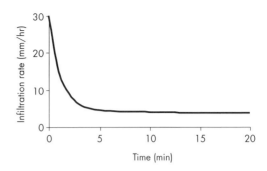

Figure 5.3 Typical infiltration curve.

Infiltration rate

The rate at which water infiltrates the soil is not constant. Generally, water initially infiltrates at a faster rate and slows down with time (see Figure 5.3). When the infiltration rate slows down to a steady level (where the curve flattens off in Figure 5.3) the **infiltration capacity** has been reached. This is the rate of infiltration when the soil is fully saturated. The terminology of infiltration capacity is misleading as it suggests a capacity value rather than a rate. In fact infiltration capacity is the infiltration rate when the water is filled to capacity with water.

Infiltration capacity is sometimes referred to as the saturated hydraulic conductivity. This is not absolutely true as the measurement is dependent on the amount of water that may be ponded on the surface creating a high hydraulic head. Saturated hydraulic conductivity should be independent of this ponded head of water. There are conditions when infiltration capacity equals saturated hydraulic conductivity, but this is not always the case.

The curve shown in Figure 5.3 is sometimes called the Philip curve, after Philip (1957) who built upon the pioneering work of Horton (1933) and provided sound theory for the infiltration of water.

The main force driving infiltration is gravity, but it may not be the only force. When soil is very dry it exerts a pulling force (soil suction; see p. 54) that will suck the infiltrating water towards the drier area. With both of these forces the infiltrating water

moves down through the soil profile in a wetting front. The wetting front is three-dimensional, as the water moves outwards as well as vertically down. The shape of the curve in Figure 5.3 is related to the speed at which the wetting front is moving. It slows down the further it gets away from the surface as it takes longer for the water at the surface to feed the front (and as the front increases in size).

Capillary forces

It is obvious to any observer that there must be forces acting against the gravitational force driving infiltration. If there were not counteractive forces all water would drain straight through to the water table leaving no water in the unsaturated zone. The counteractions are referred to as **capillary forces**. Capillary forces are actually a combination of two effects: surface tension and adsorption.

The surface tension of water is caused by the molecules in liquid water having a stronger attraction to one another than to water molecules in the air (vapour). This is due to the hydrogen bonding of water molecules described in Chapter 1. Surface tension prevents the free drainage of water from small pores within a soil by creating a force to keep the water molecules together rather than allowing them to be drawn apart. Equally, surface tension creates a force to counter the removal of water by evaporation from within the soil.

Adsorption is the force exerted through an electrostatic attraction between the faces of soil particles and water molecules. Essentially, through adsorption the water is able to stick to the surface of soil particles and not be drained away through gravity. In Chapter 1 it was pointed out that the dipolar nature of water molecules leads to hydrogen bonding (and hence surface tension). Equally the dipolar shape of water molecules lends itself to adsorptive forces.

Soil suction

Combined together, adsorption and surface tension make up the capillary force. The strength of that force is referred to as the **soil suction** or **soil moisture tension**. This reflects the concept that the capillary forces are sucking to hold onto the water and the water is under tension to keep in place. The strength of the soil suction is dependent on the amount of water present and the pore size distribution within the soil. Because of this relationship it is possible to find out pore size distribution characteristics of a soil by looking at how the soil moisture content changes at a given soil suction. This derives a *suction–moisture* or **soil moisture characteristic curve**.

A **suction–moisture curve** (see Figure 5.4) may be derived for a soil sample using a pressure plate apparatus. This increases the atmospheric pressure surrounding a soil sample and forces water out of pores and through a ceramic plate at its base. When no more water can be forced out then it is assumed that the capillary forces (i.e. the soil suction) equal the atmospheric pressure and the sample can be weighed to measure the moisture content. By steadily increasing the pressure between soil moisture measurements a suction–moisture curve can be derived. This can be interpreted to give important information on soil pore sizes and it also important for deriving an unsaturated hydraulic conductivity value for a given soil moisture (Klute, 1986).

There is a major problem in interpreting suction–moisture curves, namely **hysteresis**. In short, the water content at a given soil suction depends on whether the soil is being wetted or dried. There is a different shaped curve for wetting soils than for drying ones, a fact that can be related to the way that water enters and leaves pores. It takes a larger force for air to exit a narrow pore neck (e.g. when it is drying out) than for water to enter (wetting). Care must be taken in interpreting a suction–moisture curve, as the method of measurement may have a large influence on the overall shape.

Water in the saturated zone

Once water has infiltrated through the unsaturated zone it reaches the water table and becomes groundwater. This water moves slowly and is not available for evaporation (except through transpiration in deep-rooted plants), consequently it has a long residence time. This may be so long as to provide groundwater reserves available from more pluvial (i.e. greater precipitation) times. This can be seen in the Middle East where Saudi Arabia is able to draw on extensive 'fossil water' reserves. However, it would be wrong to think that all groundwater moves slowly; it is common for substantial movement of the water and regular replenishment during wetter months. In limestone areas the groundwater can move as underground rivers, although it may take a long time for the water to reach these conduits. In terms of surface hydrology groundwater

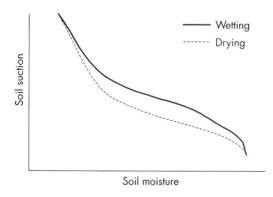

Figure 5.4 A generalised suction–moisture (or soil characteristic) curve for a soil. The two lines show the difference in measurements obtained through a wetting or drying measurement route (hysteresis)

plays an important part in sustaining streamflows during summer months.

The terminology surrounding groundwater is considerable and will not be covered in any great depth here. The emphasis is on explaining the major areas of groundwater hydrology without great detail. There are numerous texts dealing with groundwater hydrology as a separate sub-discipline, e.g. Freeze and Cherry (1979) and Price (1996).

Aquifers and aquitards

An **aquifer** is a layer of unconsolidated or consolidated rock that is able to transmit and store enough water for extraction. Aquifers range in geology from unconsolidated gravels such as the Ogallala aquifer in the USA (see Chapter 9) to distinct geological formations (e.g. chalk underlying London and much of south-east England). An **aquitard** is a geological formation that transmits water at a much slower rate than the aquifer. This is an oddly loose definition, but reflects the fact that an aquitard only becomes so relative to an aquifer. To borrow from a popular aphorism 'one man's aquifer is another man's aquitard'. The aquitard becomes so because it is confining the flow over an aquifer. In another place the same geological formation may be considered an aquifer. The term **aquifuge** is sometimes used to refer to a totally impermeable rock formation (i.e. it could never be considered an aquifer).

There are two forms of aquifers that can be seen: confined and unconfined. A *confined aquifer* has a flow boundary (aquitard) above and below it that constricts the flow of water into a confined area (see Figure 5.5). Geological formations are the most common form of confined aquifers, and as they often occur as layers the flow of water is restricted in the vertical dimension but not in the horizontal. Water within a confined aquifer is normally under pressure and if intersected by a borehole will rise up higher than the constricted boundary. If the water reaches the earth's surface it is referred to as an **artesian well**. The level that water rises up to from a confined aquifer is dependent on the amount of fall (or hydraulic head) occurring within the aquifer. This is analogous to a hose pipe acting as a syphon. If the syphon has a long vertical fall between entry of the water and exit then water will exit the hose pipe at a high velocity (i.e. under great pressure). If there is only a short vertical fall there is far less hydraulic head and the water exits at a much slower velocity. To continue the analogy further: if you could imagine that the end of the hose pipe was blocked off but that you punctured the hose, then you would expect a jet of water to shoot upwards. This jet is analogous to an artesian well.

An *unconfined aquifer* has no boundary above it and therefore the water table is free to rise and fall dependent on the amount of water contained in the aquifer (see Figure 5.6). The lower boundary of

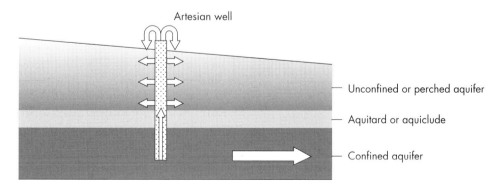

Figure 5.5 A confined aquifer. The height of water in the well will depend on the amount of pressure within the confined aquifer.

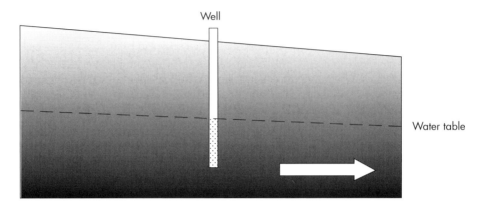

Figure 5.6 An unconfined aquifer. The water level in the well is at the water table.

the aquifer may be impervious but it is the upper boundary, or water table, that is unconfined and may intersect the surface. It is possible to have a **perched water table** or *perched aquifer* (see Figure 5.5) where an impermeable layer prevents the infiltration of water down to the regional water table. Perched water tables may be temporary features reflecting variable hydraulic conductivities within the soil and rock, or they can be permanent features reflecting the overall geology.

Groundwater flow

The movement of water within the saturated zone is described by Darcy's law. Henri Darcy was a nineteenth-century French engineer concerned with the water supply for Dijon in France. The majority of water for Dijon is aquifer fed and Darcy began a series of observations on the characteristics of flow through sand. He observed that the 'rate of flow of water through a porous medium was proportional to the hydraulic gradient' (Darcy, 1856). There are many different ways of formulating Darcy's law, but the most common and easily understood is given here:

$$Q = -K_{sat}.A.\frac{dh}{dx}$$

The **discharge** (*Q*) from an aquifer equals the saturated hydraulic conductivity (*K*) multiplied by the cross-sectional area (*A*) multiplied by the hydraulic gradient (d*h*/d*x*). The negative sign is convention based on where you measure the hydraulic gradient from (i.e. a large fall in gradient is negative).

The *h* term in the hydraulic gradient includes both the elevation and pressure head. In an unconfined aquifer it can be assumed that the hydraulic gradient is equal to the drop in height of water table over a horizontal distance (i.e. the elevation head). In a confined aquifer it is the drop in phreatic surface (i.e the level that water in boreholes reaches given the pressure the water is under) over a horizontal distance. The *h* term then includes a pressure head.

Darcy's law is an empirical law (i.e. based on experimental observation) that appears to hold under many different situations and spatial scales. It underlies most of groundwater hydrology and is very important for the management of groundwater resources. It is the term 'hydraulic conductivity' (K_{sat}) that is so important. This is the ability of a porous medium to transmit water. This can be related to the size of pores within the soil or rock and the interconnectivity between these pores. Table 5.1 shows some common values of hydraulic conductivities for soils, in addition to their porosity values.

Table 5.1 Soil hydrological properties for selected soil types. Derived from measurements of different soil types in the USA

Soil type	Saturated hydraulic conductivity (cm/hr)	Porosity
Sandy loam	2.59	0.45
Silt loam	0.68	0.50
Clay loam	0.23	0.46
Clay	0.06	0.475

Source: From Rawls *et al.* (1982)

One of the major difficulties in applying Darcy's law is that hydraulic conductivities vary spatially at both micro and macro scales. Although *K* can be measured from a small sample in the laboratory (Klute and Dirksen, 1986), in the management of water resources it is more common from larger-scale well-pumping tests (see Freeze and Cherry, 1979 for more details). The well-pumping test gives a spatially averaged K_{sat} value at the scale of interest to those concerned with water resources.

The relationship between groundwater and surface water

It is traditional to think of groundwater sustaining streamflows during the summer months, which indeed it often does. However, the interaction between groundwater and streamflow is complex and depends very much on local circumstances. Water naturally moves towards areas where faster flow is available and consequently can be drawn up towards a stream. This is the case in dry environments but is dependent on there being an unconfined aquifer near to the surface. If this is not the case then the stream may be contributing water to the ground through infiltration. Figure 5.7 shows two different circumstances of interaction between the groundwater and stream. In 5.7(a) the groundwater is contributing water to the streamflow as the water table is high. In 5.7(b) the water table is low and the stream is contributing water to the groundwater. This is commonly the case where the main river source may be mountains a considerable distance away and the river flows over an alluvial plain with the regional groundwater table considerably deeper than stream level. The interaction between groundwater and streamflow is discussed further in Chapter 6, especially with respect to stormflows.

Measuring water beneath the surface

Measurement of soil water

Gravimetric method: The simplest and most accurate means for the measurement of soil water is using the

a)

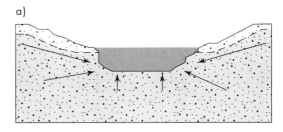

b)

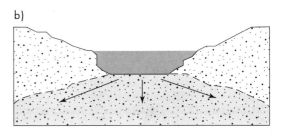

Figure 5.7 The interactions between a river and the groundwater. In (a) the groundwater is contributing to the stream, while in (b) the opposite is occurring.

gravimetric method. This involves taking a soil sample, weighing it wet, drying in an oven and then weighing it dry. Standard practice for the drying of soils is 24 hours at 105°C (Gardner, 1986). The difference between the wet and dry weights tells you how wet the soil was. If it is volumetric soil moisture content (θ) that is required then you must take a sample of known volume. This is commonly done using an undisturbed soil sampler. For gravimetric soil moisture content it does not matter what volume the sample is.

Gravimetric analysis is simple and accurate but does have several drawbacks. Most notable of these is that it is a destructive sampling method and therefore it cannot be repeated on the same soil sample. This may be a problem where there is a requirement for long-term monitoring of soil moisture. In this case a non-destructive moisture-sampling method is required. There are three methods that fit this bill, but they are indirect estimates of soil moisture rather than direct measurements as they rely on measuring other properties of soil in water. The three methods are: **neutron probes**, *electrical resistance blocks* and **time domain reflectometry**. All of these can give good results for monitoring soil moisture content, but are indirect and require calibration against the gravimetric technique.

Neutron probe: A neutron probe has a radioactive source that is lowered into an augured hole; normally the hole is kept in place as a permanent access tube using aluminium tubing. The radioactive source emits fast (or high energy) neutrons that collide with soil and water particles. The fast neutrons are very similar in size to a hydrogen ion (H^+ formed in the disassociation of the water molecule) so that when they collide the fast neutron slows down and the hydrogen ion speeds up. In contrast, when a fast neutron collides with a much larger soil particle it bounces off with very little loss of momentum. The analogy can be drawn to a pool table. When the cue ball (i.e. a fast neutron) collides with a coloured pool ball (i.e. a water particle) they both move off at similar speeds, the cue ball has

slowed down and the coloured ball sped up. In contrast, if the cue ball hits the cushion on the edge of a pool table (i.e. a soil particle) it bounces off with very little loss of speed. Consequently the more water there is in a soil the more fast neutrons would slow down to become 'slow neutrons'. A neutron probe counts the number of slow neutrons returning towards the radioactive tip, and this can be related to the soil moisture content. The neutron probe readings need to be calibrated against samples of soil with known moisture contents. This is often done by using gravimetric analysis on the samples collected while the access tubes are being put in place. It is also possible to calibrate the probe using reconstituted soil in a drum or similar vessel. It is important that the calibration occurs on the soil actually being measured as the fast:slow neutron ratio will vary according to mineralogy of the soil.

Although the neutron probe is essentially non-destructive in its measurement of soil water content, it is not continuous. There is a requirement for an operator to spend time in the field taking measurements at set intervals. This may present difficulties in the long-term monitoring of soil moisture. Another difficulty with a neutron probe is that the neutrons emitted from the radioactive tip move outwards in a spherical shape. When the probe tip is near the surface some of the sphere of neutrons will leave the soil and enter the atmosphere, distorting the reading of returning slow neutrons. A very careful calibration has to take place for near-surface readings and caution must be exercised interpreting these results. This is unfortunate as it is often the near-surface soil moisture content that is of greatest importance. Although neutron probes are reliable instruments for the monitoring of soil moisture the cost of the instruments, difficulties over installing access tubes (Figure 5.8), calibration problems and the near-surface problem have meant that they have seldom been used outside a research environment.

Electrical resistance blocks: Electrical resistance blocks use a measurement of electrical resistance to infer the water content of a soil. As water is a conductor of electricity it is reasonable to assume

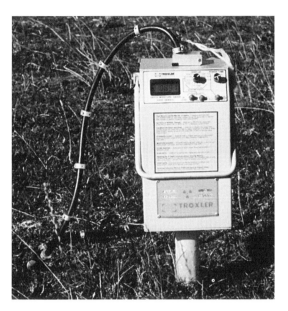

Figure 5.8 A neutron probe sitting on an access tube. The black cable extends down into the tube with the source of fast neutrons (and counter) at the tip.

that the more water there is in a soil the lower the electrical resistance, or conversely, the higher the electrical conductivity. For this instrumentation two small blocks of gypsum are inserted into the soil and a continuous measurement of electrical resistance between the blocks recorded. The measure of electrical resistance can be calibrated against gravimetric analysis of soil moisture. The continuity of measurement in electrical resistance blocks is a great advantage of the method, but there are several problems in interpreting the data. The main difficulty is that the conductivity of the water is dependent on the amount of dissolved ions contained within it. If this varies, say through the application of fertiliser, then the electrical resistance will decrease in a manner unrelated to the amount of water present. The second major difficulty is that the gypsum blocks deteriorate with time so that their electrical conductivity alters. This makes for a gradually changing signal, requiring constant recalibration. The ideal situation for the use of electrical resistance blocks is where they do not sit

in wet soil for long periods and the water moving through the soil is of relatively constant dissolved solids load. An example of this type of situation is in sand dunes, but these are not particularly representative of general land use.

Time domain reflectometry: Time domain reflectometry (TDR) is a relatively new soil moisture measurement technique. The principle of measurement is that as a wave of electromagnetic energy is passed through a soil the wave properties will alter. The way that these alter will vary, dependent on the water content of the soil. TDR measures the properties of microwaves as they are passed through a soil and relates this to the soil moisture content. Although this sounds relatively simple it is a complicated technique that requires detailed electronic technology. Up until the late 1990s this had restricted the usage of TDR to laboratory experiments but there are now soil moisture probes available that are small, robust and reliable in a field situation. An example of this is the Theta probe shown in Figure 5.9.

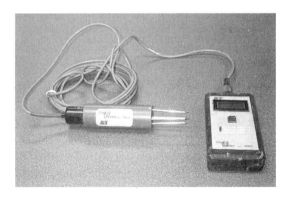

Figure 5.9 The Theta probe (manufactured by Delta-T devices); an example of a small, time domain reflectometry instrument used to measure soil moisture content in the field. The metal spikes are pushed into the soil and the moisture level surrounding them is measured.

Tensiometers

A **tensiometer** is used to measure the soil suction pressure or soil moisture tension. This is the force exerted by capillary forces and it increases as the soil dries out. A tensiometer is a small ceramic cup on the end of a sealed tube of water. The dry soil attempts to suck the water from the water-filled tube through the ceramic cup. At the top of the tube a diaphragm measures the pressure exerted by this suction. The units of soil suction are negative pressure units in that it is a negative pressure source.

Piezometers and wells

Most of the techniques described in this section have been for the measurement of water in the unsaturated zone. The main measurement techniques for water in the saturated zone are through **piezometers** and **wells**. Both of these measure the height of the water table but in slightly different ways. A *piezometer* is a tube with holes at the base that is placed at depth within a soil or rock mantle. The height of water recorded in the piezometer is thus a record of the pressure exerted by the water at the base of the tubing. A *well* has permeable sides all the way up the tubing so that water can enter or exit from anywhere up the column. Wells are commonly used for water extraction and monitoring the water table in unconfined aquifers. Piezometers are used to measure the water pressure at different depths in the soil. They are in effect the reverse of tensiometers as they are measuring a positive water pressure. In a research situation monitoring soil water dynamics in the field it is common to have a nest of piezometers and tensiometers at different levels within the soil.

Measurement of infiltration rate

Infiltration rate is measured by recording the rate at which water enters the soil. There are numerous methods available to do this, the simplest being a ring **infiltrometer** (see Figure 5.10). A solid ring is pushed into the ground and a pond of water sits on the soil (within the ring). This pond of water is kept

Figure 5.10 A single ring infiltrometer. The ring has been placed on the ground and a pond of water is maintained in the ring by the reservoir above. A bubble of air is moving up the reservoir as the water level in the pond has dropped below the bottom of the reservoir. A reading of water volume in the reservoir is taken and the time recorded.

at a steady level by a reservoir held above the ring. Recordings of the level of water in the reservoir (with time) give a record of the infiltration rate. To turn the infiltration volume into an infiltration depth the volume of water needs to be divided by the cross-sectional area of the ring.

A simple ring infiltrometer provides a measure of the ponded infiltration rate, but there are several associated problems. The first is that by using a single ring a large amount of water may escape around the sides of the ring, giving higher readings than would be obtained from a completely saturated surface. To overcome this a double ring infiltrometer is sometimes used. With this, a second wider ring is placed around the first and filled with water so that the area surrounding the measured ring is continually wet. The second problem is that ponded infiltration is a relatively rare event across a catchment. It is more common for rainfall to infiltrate directly without causing a pond to form on the surface. To overcome this a rainfall simulator may be used to provide the infiltrating water.

Estimating water beneath the surface

In the previous section it was stated that several of the methods listed were indirect measurement methods or estimation techniques. They certainly do not measure soil moisture content directly, but they have a good degree of accuracy and are good measures of soil moisture, albeit in a surrogate form. Estimating the amount of water beneath the surface can also be carried out using either numerical modelling or remote sensing techniques. The main groundwater modelling techniques focus on the movement of water in the subsurface zone, using different forms of Darcy's law. There are also models of soil water balance that rely on calculating inflows (infiltration from rainfall) and outflows (seepage and evaporation) to derive a soil water storage value for a given time and space.

A less reliable technique for the estimation of soil moisture is through remote sensing. There are three satellite remote sensing techniques of relevance to soil moisture assessment: thermal imagery, passive microwave and active microwave. All of these techniques sense the soil moisture content at the very near surface (i.e. within the top 5 cm), which is a major restriction on their application. However, this is an important area in the generation of runoff (see Chapter 6) and is still worthy of measurement.

Thermal imagery

The high heat capacity of water means that it has considerable effect on the emission of thermal

infrared signals from the earth. These can be detected by satellites and an inference made about how wet the soil is. This is especially so if two images of the same scene can be compared in order to derive a relative wetness. Satellite platforms like LANDSAT and SPOT are able to use this technique at spatial resolutions of around 10–30 m. The major difficulty with this technique is that it relies on a lack of cloud cover over the site of investigation, something that is not easy to guarantee, especially in hydrologically active (i.e. wet) areas.

Passive microwave

The earth surface emits microwaves at a very low level that can be detected by satellites. These are referred to as passive microwaves, in the sense that the earth emits them regardless of whether the satellite is present or not. The amount of water present on or above the surface of the earth affects the passive microwave signal emitted (it is a lower signal the more water there is present). The SSM/I satellite platform is able to measure passive microwaves, but only at a very coarse spatial resolution ($\approx 100 \times 100$ km). This is the major drawback of this technique at present.

Active microwave

Active microwaves are emitted from a satellite and the strength of returning signal measured. This is a complex radar system and has only been available on satellites since the early 1990s. The strength of microwave backscatter is primarily dependent on two factors: the soil wetness and the surface roughness. Where soil roughness is well known and there is little or no vegetation cover the radar backscatter has been well correlated with surface moisture (Griffiths and Wooding, 1996; Davie et al., 2001). This technique offers great hope for future estimation of soil moisture from satellites, but there is still much research to be done in understanding the role vegetation plays in affecting radar backscatter.

The great advantage of any remote sensing technique is that it automatically samples over a wide spatial area. The satellite measures the electro-magnetic radiation within each pixel; this is an average value for the whole area, rather than a point measurement that might be expected from normal soil moisture measurements. The question that needs to be answered before satellite remote sensing is widely accepted in hydrology is whether the enhanced spatial distribution of measurement is sufficient to overcome the undoubted limitations of direct soil moisture measurement.

Case study

REMOTE SENSING OF SOIL MOISTURE AS A REPLACEMENT FOR FIELDWORK

Remote sensing of soil moisture may offer a way of deriving important hydrological information without intensive, and costly, fieldwork programmes. Grayson *et al.* (1992) suggest that this could be used to set the initial conditions for hydrological modelling, normally a huge logistical task. One of the major difficulties in this is that the accuracy of information derived from satellite remote sensing is not high enough for use in hydrological modelling. As a counter to this it can be argued that the spatial discretisation offered by remote sensing measurements is far better than that available through traditional field measurement techniques.

In an attempt to reconcile these differences Davie *et al.* (2001) intensively monitored a 15-hectare field in eastern England and then analysed the satellite-derived, active micro-wave backscatter for the same period. The field programme consisted of measuring surface soil moisture (gravimetric method) at three different scales in an attempt to spatially characterise the soil moisture. The three scales consisted of: (a) thirteen samples taken 1 m apart; (b) lines of samples 30 m apart; and (c) the lines were approximately 100 m apart. The satellite data was from the European Remote Sensing Satellite (ERS) using **Synthetic Aperture Radar (SAR)**.

Analysis of the field data showed great variation in the surface soil moisture and a difference in measurement depending on the scale of measure-ment (see Figure 5.11). It is evident from Figure 5.11 that the point measurements of soil moisture are highly variable and that many measurements of soil moisture are required to try and characterise the overall field surface soil moisture.

In contrast to the directly measured soil mois-ture measurements, the microwave backscatter shows much less variation in response (see figure 3, p. 330 of Davie *et al.*, 2001). The most likely explanation for this is that the spatial averaging of backscatter response within a pixel (25 × 25m in this case) evens out the variations found through point measurements. This can be thought of as a positive aspect because in many cases for hydro-logical modelling the scale needed is larger than point measurements and the spatial integration of

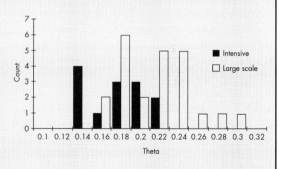

Figure 5.11 Measured surface soil moisture distributions at two different scales for a field in eastern England in October 1995.

backscatter response may be a way of getting around highly variable point measurements. When further analysis of the backscatter response was carried out it showed that interpretation at the pixel size was meaningless and they needed to be averaged-up themselves. It was found that the smallest resolution to yield meaningful results was around 1 hectare (100 × 100 m).

As with other studies on bare fields (e.g. Griffiths and Wooding, 1996) there was a reasonably good correlation between radar backscatter and measured soil moisture. Unfortunately, this relationship is not good enough to provide more than around 70 per cent accuracy on estimations

of soil moisture. In addition to this the study was carried out in conditions ideally suited for SAR interpretation (flat topography with no vegetation cover) which are far from atypical conditions encountered in hydrological investigations.

Overall it is possible to say that although the satellite remote sensing of soil moisture using SAR may offer some advantages of spatial integration of the data it is not enough to offset the inaccuracy of estimation, particularly in non-flat, vegetated catchments. There may come a time when satellite remote sensing can be used as a replacement for field measurements, but it is in the future.

SNOW AND ICE

Snow and ice are hugely important stores of water for many countries in the world, particularly at high latitudes or where there are large mountain ranges. The gradual release of water from snow and ice, either during spring and summer or on reaching a lower elevation, makes a significant impact on the hydrology of many river systems.

The timing of snow- and ice-melt is critical in many river systems, but especially so in rivers that flow north towards the Arctic Circle. In this case the melting of snow and ice may occur in the headwaters of a river before it has cleared further downstream (at higher latitudes). This may lead to a build up of water behind the snow and ice further downstream – a snow and ice dam (see Plate 5). Beltaos (2000) estimates that the cost of damage caused by ice-jams in Canada is around $60 million dollars per annum. The Case Study outlined on p. 64 gives an example of the flooding that may result from this type of snow-melt event.

Measuring snow depth

The simplest method of measuring snow depth is the use of a core sampler. This takes a core of snow, recording its depth. The snow sample can then be

melted to derive the water equivalent depth, the measurement of most importance in hydrology. The major drawbacks in using a core sampler to derive snow depth are that it is a non-continuous reading (similar to daily rainfall measurement) and the position of coring may be critical (because of snow drifting).

A second method of measuring snow depth is to use a *snow pillow*. This is a sealed plastic pillow that is normally filled with some form of antifreeze and connected to a pressure transducer. When left out over a winter period the weight of snow on top of the pillow is recorded as an increasing pressure, which can be recalculated into a mass of snow. It is important that the snow pillow does not create an obstacle to drifting snow in its own right. To overcome this, and to lessen the impact of freezing on the pillow liquid, it is often buried under a shallow layer of soil or laid flat on the ground. When connected to a continuous logging device a snow pillow provides the best record of snow depth (and water equivalent mass).

Estimating snow cover

The main method of estimating snow cover is using satellite remote sensing. Techniques exist that give a reasonably good method of detecting the areal

Case study

THE MACKENZIE RIVER: DEMONSTRATING THE INFLUENCE OF SNOW AND ICE ON RIVER FLOWS

The Mackenzie river is at the end of one of the great river systems of the world. In North America the Missouri/Mississippi system drains south, the Great Lakes/Hudson system drains eastwards, and the Mackenzie river system drains northwards with its mouth in the Beaufort Sea (part of the Arctic Ocean). The Mackenzie river itself has a length of 1,800 km from its source: the Great Slave Lake. The Peace and Athabasca rivers which flow into the Great Slave Lake, and are therefore part of the total Mackenzie drainage basin, begin in the Rocky Mountains 1,000 km to the south-west. The total drainage basin is approximately 1,841,000 km^2, making it the twelfth largest in the world, and the river length is 4,250 km (see Figure 5.12).

The high latitude of the Mackenzie river makes for a large component of snow and ice melt within the annual hydrograph. This can be seen in Figure 5.13, taken at the junction of the Mackenzie and Arctic Red rivers. This gauging station is well within the Arctic circle and towards the delta of the Mackenzie river. The monthly discharge values increase dramatically from April to June when the main melt occurs and then gradually decrease to reach a minimum value during the winter months. The highest average streamflow occurs in June despite the highest precipitation occurring one

month later in July. Overall there is very little variation in precipitation but a huge variation on riverflow. This an excellent example of the storage capability of snow and ice within a river catchment. Any water falling during the winter months is trapped in a solid form (snow or ice) and may be released only during the warmer summer months. The amount of precipitation falling during the summer months (mostly rainfall) is dwarfed by the amount of water released in the melt during May and June.

The most remarkable feature of a river system such as the Mackenzie is that the melt starts occurring in the upper reaches, sending a pulse of water down the river, before ice on the lower reaches has melted properly. This is not unique to the Mackenzie river, all the great rivers draining northwards in Europe, Asia and North America exhibit the same tendencies. If we look in detail at an individual year (Figure 5.14) you can see the difference in daily hydrographs for stations moving down the river (i.e. northwards). Table 5.2 summarises the information on latitude and flow characteristics for the Mackenzie. It is not a simple story to decipher (as is often the case in hydrology), but you can clearly see that the rise in discharge at the Arctic Red River station starts later than the Norman Wells station further to the

Table 5.2 Summary of latitude and hydrological characteristics for three gauging stations on the Mackenzie river

Mackenzie river gauging station	Latitude (north)	Date of last ice on river (1995)	Date of peak discharge (1995)
Fort Simpson	61° 52′ 7″	14 May	21 May
Norman Wells	65° 16′ 26″	18 May	10 May
Arctic Red River	67° 27′ 30″	31 May	14 May

Source: Data courtesy of Environment Canada

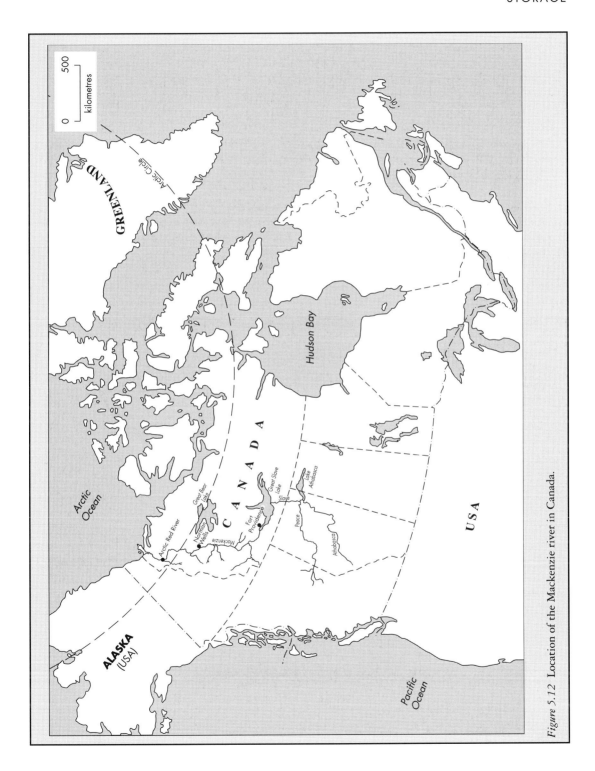

Figure 5.12 Location of the Mackenzie river in Canada.

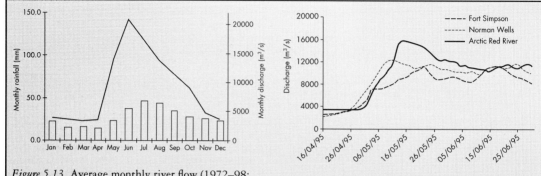

Figure 5.13 Average monthly river flow (1972–98; line) for the Mackenzie River at the Arctic Red River gauging station (latitude 67° 27' 30" north) and average precipitation (1950–94) for the Mackenzie river basin (bars).
Source: Data courtesy of Environment Canada

Figure 5.14 Daily river flow at three locations on the Mackenzie river from mid-April through to the end of June 1995.
Source: Data courtesy of Environment Canada

south. The rise is caused by melt, but predominantly from upstream. It is also clear that for both Norman Wells and the Arctic Red River stations the highest discharge value of the year is occurring while the river is still covered in ice. This creates huge problems for the drainage of the area as the water may build up behind an ice dam. Certainly the water flowing under the ice will be moving much quicker than the ice and water mix at the surface. Plate 5 demonstrates the way that water builds up behind an ice dam, particularly where there is a constriction on either side of the river channel.

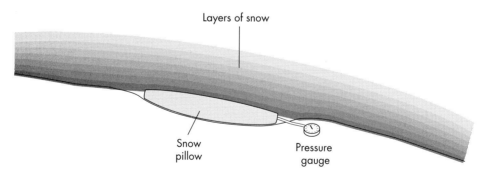

Figure 5.15 Snow pillow for measuring the depth of snow above a point. The snow depth is recorded as pressure exerted on the pillow.

extent of snow cover, but it is much harder to translate this into a volume of snow (i.e. by knowing depth) or water. Optical and thermal infrared data can be used to estimate snow cover but they rely heavily on the reflective ability of snow; unfortunately, other surfaces such as clouds may also exhibit these properties (Fitzharris and McAlevey, 1999). Microwave data offer a far better method of detecting snow cover. Passive microwaves detected by a satellite can be interpreted to give snow cover because any water (or snow) covering the surface absorbs some of the microwaves emitted by the earth surface. The greater the amount of snow the weaker is the microwave signal received by the satellite. Ranzi *et al.* (1999) have used AVHRR imagery to monitor the snowpack in an area of northern Italy and some of Switzerland and compared this to other measurement techniques. Unfortunately, the passive microwave satellite sensors currently available (e.g. AVHRR) are at an extremely coarse spatial resolution that is really only applicable at the large catchment scale. Active microwave sensing offers more hope but its usage for detecting snow cover is still being developed.

Snow-melt

Of critical importance to hydrology is the timing of snow-melt, as this is when the stored water is becoming available water. There are numerous models that have been developed to try and estimate the amount of snow-melt that will occur. Ferguson (1999) gives a summary of recent snow-melt modelling work. The models can be loosely divided into those that rely on air temperature and those that rely on the amount of radiation at a surface. The former frequently use a degree days approach, the difference between mean daily temperature and a melting threshold temperature. Although it would seem sensible to treat zero as the melting threshold temperature this is not always the case, snow will melt with the air temperature less than zero because of energy available through the soil (soil heat flux) and solar radiation. The degree day snow-melt approach calibrates the amount of snow that might be expected given a certain value of degree day. Although this is useful for hydrological studies it is often difficult to calibrate the model without detailed snow-melt data.

SUMMARY

Water held in storage is an important part of the water balance equation. It is of particular importance as a change in storage, whether as an absorption term (negative) or a release (positive). As with all the processes in the hydrological cycle, storage is difficult to measure accurately at a useful spatial scale. This applies whether it is water held underground or as snow and ice. The release of water from storage may have a significant effect on river flows, as is demonstrated in the Makenzie river Case Study (p. 64).

ESSAY QUESTIONS

1   **Compare and contrast different methods for measuring soil water content at the hillslope scale.**

2   **Define the term *saturated hydraulic conductivity* and explain its importance in understanding groundwater flow.**

3   **Explain the terms *confined* and *unconfined* with respect to aquifers and describe how artesian wells come about.**

4   **Describe the field experiment required to assess the amount of snow in a seasonal snowpack and the timing of the snow-melt.**

FURTHER READING

Ferguson, R.I. (1999) Snowmelt runoff models. *Progress in Physical Geography* 23:205–228.
An overview of snow-melt estimation techniques.

Freeze, R.A. and Cherry J.A. (1979) *Groundwater.* Prentice-Hall, Englewood Cliffs, N.J.
A classic text book on groundwater (including soil water).

Klute, A. (ed.) (1986) *Methods of soil analysis. Part 1: Physical and mineralogical methods.* American Society

of Agronomy–Soil Science Society of America, Madison, Wisc.
A mine of information on soil methods.

Price, M. (1996) *Introducing groundwater* (2nd edn). Chapman and Hall, London.
An introductory text on groundwater.

6

RUNOFF

LEARNING OBJECTIVES

When you have finished reading this chapter you should have:

- An understanding of the process of runoff leading to **channel flow**.
- A knowledge of the techniques for measuring streamflow and runoff directly.
- A knowledge of techniques used to estimate streamflow.
- A knowledge of how numerical models are used in hydrology to estimate streamflow and runoff.

The amount of water within a river is of great interest to hydrologists. It represents the end-product of all the other processes and is where the largest amount of effort has gone into analysis of historical records. The methods of analysis are covered in Chapter 7; this chapter deals with the mechanisms that lead to water entering the stream: the runoff mechanisms. *Runoff* is a loose term that covers the movement of water to a channelised stream, after it has reached the ground as precipitation. The movement can occur either on or below the surface and at differing velocities. Once the water reaches a stream it moves towards the oceans in a channelised form, the process referred to as *streamflow* or *riverflow*. Another term for streamflow is *discharge*. This is the volume of water plotted against time and the SI units are m^3/s (*cumecs*). A continuous record of streamflow is called a **hydrograph** (see Figure 6.1). Although we think of this as continuous measurement it is normally either an averaged flow over a time period or a series of samples (e.g. hourly records).

In Figure 6.1 there are a series of peaks between periods of steady, much lower flows. The hydrograph peaks are referred to as **peakflow**, **stormflow** or even **quickflow**. They are the water in the stream during and immediately after a storm event. The steady periods between peaks are referred to as *baseflow* or sometimes **slowflow** (NB This is different from **low flow**; see Chapter 7). The shape of a hydrograph, and in particular the shape of the stormflow peak, is influenced by the storm characteristics and many

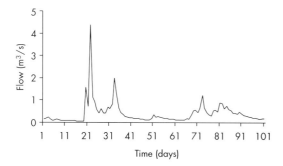

Figure 6.1 A typical hydrograph, taken from the river Wye, Wales for a 100-day period during the autumn of 1995. The values plotted against time are mean daily flow in cumecs.

different factors within a catchment. The largest influence is exerted by catchment size, but other factors include slope angles, shape of catchment, soil type, vegetation type and percentage cover, degree of urbanisation and the antecedent soil moisture.

Figure 6.2 shows the shape of a storm hydrograph in detail. There are several important hydrological terms that can be seen in this diagram. The **rising limb** of the hydrograph is the initial steep part leading up to the highest or peakflow value. The water contributing to this part of the hydrograph is from *channel precipitation* (i.e. rain that falls directly onto the channel) and rapid runoff mechanisms. Some texts claim that channel precipitation shows up as a preliminary blip before the main rising limb. In reality this is very rarely observed, a factor of the

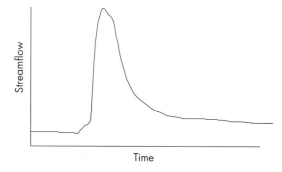

Figure 6.2 Demonstration storm hydrograph.

complicated nature of storm runoff processes. The **recession limb** of the hydrograph is after the peak and is characterised by a long, slow decrease in streamflow until the baseflow is reached again. The recession limb is attenuated by two factors: storm water arriving at the mouth of a catchment from the furthest parts, and the arrival of water that has moved as underground flow at a slower rate than the streamflow.

Exactly how water moves from precipitation reaching the ground surface to channelised streamflow is one of the most intriguing hydrological questions, and one that cannot be answered easily. Much research effort in the past hundred years has gone into understanding runoff mechanisms; considerable advances have been made, but there are still many unanswered questions. The following section describes how it is believed runoff occurs, but there are many different scales at which these mechanisms are evident.

RUNOFF MECHANISMS

Figure 6.3 is an attempt to represent the different runoff processes that can be observed at the hillslope scale. **Overland flow** (Q_o) is that which runs across the surface of the land before reaching the stream. In the subsurface, throughflow (Q_t) (some authors refer to this as **lateral flow**) occurs in the shallow subsurface, predominantly, although not always, in the unsaturated zone. Groundwater flow (Q_G) is in the deeper saturated zone. All of these are runoff mechanisms that contribute to streamflow. The relative importance of each is dependent on the catchment under study and the rainfall characteristics during a storm.

Overland flow

Some of the earliest research work on how overland flow occurs was undertaken by Horton, an engineer working predominantly in the American Midwest. Horton (1933) hypothesised that overland flow occurred when the rainfall rate was higher than the

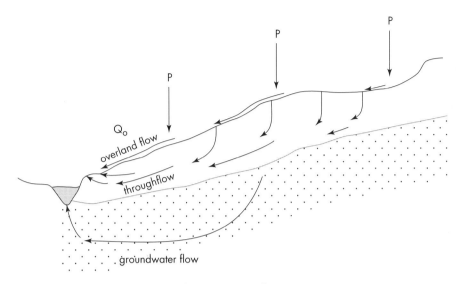

Figure 6.3 Hillslope runoff processes. See text for explanation of terms.
Source: Adapted from Dunne (1978)

infiltration rate of a soil. Horton went on to suggest that under these circumstances the excess rainfall collected on the surface before travelling towards the stream as a thin sheet of water moving across the surface. Under this hypothesis it is the infiltration rate of a soil that acts as a controlling barrier or partitioning device. Where the infiltration capacity of a soil is low, overland flow occurs readily. This type of overland flow is referred to as **infiltration excess overland flow** or **Hortonian overland flow**.

Horton's ideas were extremely important in hydrology as they represented the first serious attempt to understand the processes of storm runoff that lead to a storm hydrograph. Unfortunately, when people started to measure infiltration rates of soils they invariably found that they were far higher than most normal rainfall rates. This is illustrated in Table 6.1 where some typical infiltration and rainfall rates are shown. In addition to this it is extremely rare to find a thin sheet of water moving over the surface during a storm event. It was observations such as those by Hursh (1944) and others that led to a general revision of Horton's hypothesis. Betson (1964) proposed the idea that within a catchment there are only limited areas that contribute

Table 6.1 Some typical infiltration rates compared to rainfall intensities

Soil and vegetation	Infiltration rate (mm/hr)	Rainfall type	Rainfall intensity (mm/hr)
Forested loam	100–200	Thunderstorm	50–100
Loam pasture	10–70	Heavy rain	5–20
Sand	3–15	Moderate rain	0.5–5
Bare clay	0–4	Light rain	0.5

Source: From Burt (1987)

overland flow to a storm hydrograph. This is referred to as the **partial areas concept**. Betson did not challenge the role of infiltration excess overland flow as the primary source of stormflow, but did challenge the idea of overland flow occurring as a thin sheet of water throughout a catchment.

Hewlett and Hibbert (1967) were the first to suggest that there might be another mechanism of overland flow occurring. This was particularly based on the observations from the eastern USA: that during a storm it was common to find all the rainfall infiltrating a soil. Hewlett and Hibbert (1967) hypothesised that during a rainfall event all the water infiltrated the surface. Through a mixture of infiltration and throughflow the water table would rise until in some places it reached the surface. At this stage overland flow occurs as a mixture of return flow (i.e. water that has been beneath the ground but returns to the surface) and rainfall falling on saturated areas. This type of overland flow is referred to as **saturated overland flow**. Hewlett and Hibbert (1967) suggested that the water table was closest to the surface, and therefore likely to rise to the surface quickest, adjacent to stream channels and at the base of slopes. Their ideas on stormflow were that the areas contributing water to the hydrograph peak were the saturated zones, and that these vary from storm to storm. The saturated areas immediately adjacent to the stream act as extended channel networks. This is referred to as the **variable source areas concept**. This goes a step beyond the ideas of Betson (1964) as the catchment has a partial areas response but the response area is dynamic and variable in space and time.

So who was right: Horton, or Hewlett and Hibbert? The answer is that both were. Table 6.2 provides a summary of the ideas for storm runoff generation described here. It is now accepted that saturated overland flow (Hewlett and Hibbert) is the dominant overland flow mechanism in humid mid-latitude areas. It is also accepted that the variable source areas concept is the most valid description of stormflow processes. However, where the infiltration capacity of a soil is low or the rainfall rates are high, Hortonian overland flow does occur. In Table 6.1 it can be seen that there are times when rainfall intensities will exceed infiltration rates under natural circumstances.

Examples of low infiltration rates can be found with compacted soils (e.g. from vehicle movements in an agricultural field), on roads and paved areas, on heavily crusted soils and what are referred to as hydrophobic soils. The latter examples are often found in arid and semi-arid regions such as the American Midwest where Horton worked. Hydrophobic soils occur where there is a mineral element in the soil that swells up on contact with water. This creates an impermeable barrier to infiltrating water, leading to Hortonian overland flow. Equally, in arid and semi-arid zones it is not uncommon to find extremely high rainfall rates (fed by convective storms) that can lead to infiltration excess overland flow and rapid flood events; this is called **flash flooding**.

In Hewlett and Hibbert's (1967) original hypothesis it was suggested that contributing saturated areas would be immediately adjacent to stream channels. Subsequent work by the likes of Dunne

Table 6.2 A summary of the ideas on how stormflow is generated in a catchment

	Horton	Betson	Hewlett and Hibbert
Infiltration	Controls overland flow	Controls overland flow	All rainfall infiltrates
Overland flow mechanism	Infiltration excess	Infiltration excess	Saturated overland flow
Contributing area	Uniform throughout the catchment	Restricted to certain areas of the catchment	Contributing area is variable in time and space

and Black (1970), Anderson and Burt (1978) and others has identified other areas in a catchment prone to inducing saturated overland flow. These include hillslope hollows, slope concavities (in section) and where there is a thinning of the soil overlying an impermeable base. In these situations any throughflow is likely to return to the surface as the volume of soil receiving it is not large enough for the amount of water entering it. This can be commonly observed in the field where wet and boggy areas can be found at the base of slopes and at valley heads (hillslope hollows).

Subsurface flow

Under the variable source areas concept there are areas within a catchment that contribute overland flow to the storm hydrograph. When we total up the amount of water found in a storm hydrograph it is difficult to believe that it has all come from overland flow, especially when this is confined to a relatively small part of the catchment. The manner in which the recession limb of a hydrograph attenuates the stormflow suggests that it may be derived from a slower movement of water: subsurface flow. In addition to this, tracer studies looking at where the water has been before entering the stream as stormflow have found that a large amount of the storm hydrograph consists of 'old water' (e.g. Martinec *et al.*, 1974; Fritz *et al.*, 1976). This old water has been sitting in the soil, or as fully saturated groundwater, for a considerable length of time and yet enters the stream during a storm event. There have been several theories put forward to try and explain these findings, almost all involving throughflow.

Throughflow is a general term used to describe the movement of water through the unsaturated zone, normally this is the soil matrix. One of the first theories put forward concerning the contribution of throughflow to a storm hydrograph was by Horton and Hawkins (1965) (this Horton being a different person from the proposer of Hortonian overland flow), followed by Hewlett and Hibbert (1967). They proposed the mechanism of *translatory* or *piston*

flow to explain the rapid movement of water from the subsurface to the stream. They suggested that as water entered the top of a soil column it displaced the water at the bottom of the column (i.e. old water), this latter entering the stream. The analogy is drawn to a piston where pressure at the top of the piston chamber leads to a release of pressure at the bottom. Although this is a simple analogy there are considerable differences in a field situation. First and foremost a hillslope or soil column is not bounded by impermeable sides as a piston chamber is. This would suggest that with the addition of water at the top of the hillslope or soil column there would be an accommodation of that water somewhere in between the top and bottom and very little piston displacement. The accommodation would occur through the water being absorbed into drier soil above or forced onto the surface. Very few field studies have been able to show a significant role for translatory or piston flow as a contributor to stormflow. Brammer and McDonnell (1996) suggest that this may be a mechanism for the rapid movement of water along the bedrock and soil interface on the steep catchment of Maimai in New Zealand. In this case it is the hydraulic gradient created by an addition of water to the bottom of the soil column, already close to saturated, that forces water along the base where hydraulic conductivities are higher.

Ward (1984) draws the analogy of a thatched roof to describe the contribution of subsurface flow to a stream (based on the ideas of Zaslavsky and Sinai, 1981). When straw is placed on a sloping roof it is very efficient at moving water to the bottom of the roof (the guttering being analogous to a stream) without visible overland flow. This is due to the preferential flow direction along, rather than between, sloping straws. Measurements of hillslope soil properties do show a higher hydraulic conductivity in the downslope rather than vertical direction. This would account for a movement of water downslope as throughflow, but it is still bound up in the soil matrix and reasonably slow.

There is considerable debate on the role of **macropores** in the rapid movement of water through the soil matrix. Macropores are larger pores within a soil matrix, typically with a diameter greater than 3 mm. They may be caused by soils cracking, worms burrowing or other biotic activities. The main interest in them from a hydrologic point of view is that they provide a rapid conduit for the movement of water through a soil. The main area of contention concerning macropores is whether they form continuous networks allowing rapid movement of water down a slope or not. There have been studies suggesting macropores as a major mechanism contributing water to stormflow (e.g. Mosley, 1979, 1982; Wilson *et al.*, 1990), but it is difficult to detect whether these are from small areas on a hillslope or continuous throughout. Jones (1981) and Tanaka (1992) summarise the role of pipe networks (a form of continuous macropores) in **hillslope hydrology**. Where found, pipe networks have considerable effect on the subsurface hydrology but they are not a common occurrence in the field situation.

The role of macropores in runoff generation is unclear. Although they are capable of allowing rapid movement of water towards a stream channel there is little evidence of networks of macropores moving large quantities of water in a continuous fashion. Where macropores are known to have a significant role is in the rapid movement of water to the saturated layer (e.g. Heppell *et al.*, 1999).

Groundwater contribution to stormflow

Another possible explanation for the presence of old water in a storm hydrograph is that it comes from the saturated zone (groundwater) rather than from throughflow. This is contrary to conventional hydrological wisdom which suggests that groundwater contributes to baseflow but not to the stormflow component of a hydrograph. Although a groundwater contribution to stormflow had been suggested before, it was not until Sklash and Farvolden (1979) provided a theoretical mechanism for this to occur that the idea was seriously considered. They proposed the capillary fringe hypothesis to explain

the groundwater ridge, a rise in the water table immediately adjacent to a stream (as observed by Ragan, 1968). Their hypothesis suggests that the addition of a small amount of infiltrating rainfall to the zone immediately adjacent to a stream causes the soil water to move from an unsaturated state (i.e. under tension) to a saturated state (i.e. a positive pore pressure expelling water). As explained in Chapter 5, the relationship between soil water content and soil water tension is non-linear. The addition of a small amount of water can cause a rapid change in soil moisture status from unsaturated to saturated. This provides the groundwater ridge which

> not only provides the early increased impetus for the displacement of the groundwater already in a discharge position, but it also results in an increase in the size of the groundwater discharge area which is essential in producing large groundwater contributions to the stream.
>
> (Sklash and Farvolden, 1979: 65)

An important point to stress from the capillary fringe hypothesis is that the groundwater ridge is developing well before any throughflow may have been received from the contributing hillslope areas. These ideas confirm the variable source areas concept and provide a mechanism for a significant old water contribution to storm hydrographs. Field studies such as that by McDonnell (1990) have observed groundwater ridging to a limited extent, although it is not an easy task as often the instrument response time is too slow to detect the rapid change in pore pressure properly.

Summary of storm runoff mechanisms

The mechanisms that lead to a storm hydrograph are extremely complex and still not fully understood. Although this would appear to be a major failing in a science that is concerned with the distribution of water on the surface of the earth and its movement over and beneath the surface, it is also an acknowledgement of the extreme diversity found in nature. In general there is a reasonable understanding of possible storm runoff mechanisms but they are not able to be universally applied.

In some field situations the role of throughflow is important, in others not; likewise for groundwater contributions, overland flow and **pipeflow**. The challenge for modern hydrology is to identify quickly the dominant mechanisms for a particular hillslope or catchment so that the understanding of the hydrological processes in that situation can be used to aid management of the catchment.

The processes of storm runoff generation described here are mostly observable at the hillslope scale. At the catchment scale (and particularly for large river basins) the timing of peak flow (and consequently the shape of the storm hydrograph) is influenced more by the channel drainage network and the precipitation characteristics of a storm than by the mechanisms of runoff. This is a good example of the problem of scale described in Chapter 1. At the small hillslope scale storm runoff generation mechanisms are important, but they become considerably less so at the much larger catchment scale.

Baseflow

In sharp contrast to the storm runoff debate, there is general consensus that the major source of baseflow is groundwater – and to a lesser extent throughflow. This is water that has infiltrated the soil surface and moved towards the saturated zone. Once in the saturated zone it moves downslope, often towards a stream. A stream or lake is often thought to occur where the regional water table intersects the surface, although this may not always be the case. In Chapter 5 the relationship between groundwater and streamflow has been explained (see Figure 5.7).

Channel flow

Once water reaches the stream it will flow through a channel network to the main river. The controls over the rate of flow of water in a channel are to do with the volume of water present, the gradient of the channel, and the resistance to flow experienced at the channel bed. This relationship is described in uniform flow formulae such as the Chezy and Manning equations (see p. 83). The resistance to flow is governed by the character of the bed surface. Boulders and vegetation will create a large amount of friction, slowing the water down as it passes over the bed.

In many areas of the world the channel network is highly variable in time and space. Small channels may be ephemeral and in arid regions will frequently only flow during flood events. The resistance to flow under these circumstances is complicated by the infiltration that will be occurring at the water front and bed surface. Under a continual flow regime the infiltration from the stream to ground will depend on the hydraulic gradient and the infiltration capacity. With an ephemeral stream the first flush of water will infiltrate at a much higher rate as it fills the available pore space in the soil/rock at the bed surface. This will remove water from the stream and also slow the water front down as it creates a greater friction surface.

MEASURING STREAMFLOW AND HILLSLOPE RUNOFF

The techniques and research into the measurement of streamflow are referred to as **hydrometry**. Streamflow measurement can be subdivided into two important subsections: instantaneous and continuous techniques.

The measurement of runoff may be required to assess the relative contribution of different hillslope runoff processes (i.e. throughflow, overland flow, etc.). There are no standard methods for the measurement of runoff processes, different researchers use different techniques according to the field conditions expected and personal preference.

Instantaneous streamflow measurement

Velocity–area method

Streamflow or discharge is a volume of water per unit of time. The standard units for measurement of discharge are m³/s (cubic metres per second or *cumecs*). If we rewrite the units of discharge we can think of them as a water velocity (m/s) passing through a cross-sectional area (m²).

$$m^3/s = m/s \times m^2$$

The **velocity–area method** measures the stream velocity, the stream cross-sectional area and multiplies the two together. In practice this is carried out by dividing the stream into small sections and measuring the velocity of flow going through each cross-sectional area and applying the following formula:

$$Q = v_1 a_1 + v_2 a_2 + ... v_i a_i$$

where Q is the streamflow or discharge (m³/s), v is the velocity measured in each trapezoidal cross-sectional area (see Figure 6.4), and a is the area of the trapezoid (usually estimated as the average of two depths divided by the width between).

The number of cross-sectional areas that are used in a stream depend upon the width and smoothness of bed. If the stream bed is particularly rough it is necessary to use more cross-sectional areas so that the estimates are as close to reality as possible (note the discrepancy between the dashed and solid lines in Figure 6.4).

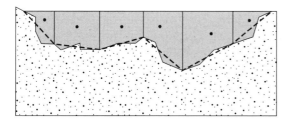

Figure 6.4 The velocity–area method of streamflow measurement. The black circles indicate the position of velocity readings. Dashed lines represent the triangular or trapezoidal cross-sectional area through which the velocity is measured.

The velocity measurement is usually taken with a flow meter. This is a propeller inserted into the stream which records the number of propeller turns with time. This reading can be easily converted into a stream velocity. In the velocity–area method it is necessary to assume that the velocity measurement is representative of all the velocities throughout the cross-sectional area. It is not normally possible to take multiple measurements so an allowance has to be made for the fact that the water travels faster along the surface than nearer the stream bed as a result of friction exerted by the stream bed. As a general rule of thumb the sampling depth should be 60 per cent of the stream depth – that is, in a stream that is 1 m deep the sampling point should be 0.6 m below the surface or 0.4 m above the bed. In a deep river it is good practice to take two measurements (one at 20 per cent and the other at 80 per cent of depth) and average the two.

Where there is no velocity meter available it may be possible to make a very rough estimate of stream velocity using a float in the stream (i.e. the time it takes to cover a measured distance). When using this method allowance must be made for the fact that the float is travelling on the surface of the stream at a faster rate than water closer to the stream bed.

The velocity–area method is an effective technique for measuring streamflow in small rivers, but its reliability is heavily dependent on the sampling strategy. The technique is also less reliable in small, turbid streams with a rough bed (e.g. mountain streams). Under these circumstance other streamflow measurement methods such as **dilution gauging** may be more applicable.

Dilution gauging

Dilution gauging works on the principle that the more water there is in a river the more it will dilute a tracer. There is a well-established relationship between the amount of the tracer found naturally in the stream (C_o), the concentration of tracer put into the river (C_i), the concentration of tracer measured downstream after mixing (C_d), and the stream discharge (Q). The type of tracer used is dependent on the equipment available; the main point is that it must be detectable in solution and non-harmful to the aquatic flora and fauna. A simple tracer that is often used is a solution of table salt (NaCl), a conductivity meter being employed to detect the salt solution.

There are two different ways of carrying out dilution gauging that use slightly different equations. The first puts a known volume of tracer into the river and measures the concentration of the 'slug' of tracer as it passes by the measurement point. This is referred to as gulp dilution gauging. The equation for calculating flow by this method is shown below:

$$Q = \frac{C_t V}{\sum (C_d - C_0)\Delta t}$$

where Q is the unknown streamflow, C is the concentration of tracer either in the slug (t), downstream (d), or background in the stream (0); Δt is the time interval. The denominator of this equation is the sum of measured concentrations of tracer downstream.

The second method uses a continuous injection of tracer into the river and measures the concentration downstream. The continuous injection method is better than the slug injection method as it measures the concentration over a greater length of time. Using the formula listed below the stream discharge can be calculated.

$$Q = q\frac{C_t - C_d}{C_d - C_0}$$

where q is the flow rate of the injected tracer (i.e. injection rate) and all other terms are as for the gulp injection method.

Probably the most difficult part of dilution gauging is calculating the distance downstream between where the tracer is injected and the river concentration measuring point (the mixing distance). This can be estimated as follows:

$$L = 0.13 C_Z \left(\frac{0.7C_Z + w}{g} \right)\left(\frac{w^2}{d} \right)$$

where L = mixing distance (m)
C_Z = Chezy's **roughness coefficient** (see Table 6.3)
w = average stream width (m)
g = gravity constant (≈ 9.8 m/s^2)
d = average depth of flow (m)

Continuous streamflow measurement

The methods of instantaneous streamflow measurement described above only allow a single measurement to be taken at a location. Although this can be repeated at a future date it requires a continuous measurement technique to give the data for a hydrograph. There are three different techniques that can be used for this method: stage discharge relationships, flumes and weirs, and ultrasonic flow gauging.

Table 6.3 Chezy roughness coefficients for some typical streams

Type of channel	Chezy roughness coefficient for a hydraulic radius of 1 m
Artificial concrete channel	71
Excavated gravel channel	40
Clean regular natural channel <30 m wide	33
Natural channel with some weeds or stones <30 m wide	29
Natural channel with sluggish weedy pools <30 m wide	14
Mountain streams with boulders	20
Streams larger than 30 m wide	40

Source: Adapted from Richards (1982)

Stage vs. discharge relationship

River **stage** is another term for the water level or height. In many cases it is possible to draw a relationship between river stage and discharge: the so-called **rating curve**. An example of a rating curve is shown in Figure 6.5. This has the advantage of allowing continuous measurement of river stage (a relatively simple task) that can then be equated to the actual discharge. The stage discharge relationship is normally derived through a series of velocity–area measurements at a particular site while at the same time recording the stage with a stilling well (see Figure 6.6). As can be seen in Figure 6.5, the rating curve is non-linear, a reflection of the river bank profile. As the river fills up between banks it takes a greater amount of water to cause a change in stage than at low levels.

The relationship between stage and discharge is dependent on frequent measurement of river discharge, normally using a point measurement technique, and assumes that the river bed profile remains static. The assumption of a static river bed profile can sometimes be problematic, leading to the installation of a concrete structure to maintain stability. One of the difficulties with the stage discharge relationship is that frequent measurements of river discharge lead to many measurements taken during periods of low and medium flow but very few at high runoff (i.e. during flood events). The lack of

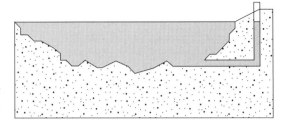

Figure 6.6 Stilling well to provide a continuous measurement of river stage (height). The height of water is measured in the well immediately adjacent to the river.

data at the extreme end of the stage vs. discharge curve may lead to difficulties in interpreting data during peak flows. It is extremely important that discharge measurements are taken during floods, but this is frequently difficult due to the hazardous nature of the flood event. When interpreting data derived from this type of streamflow measurement it is important that the hydrologist bears in mind that it is stream stage that is being investigated and that from this the stream discharge is inferred.

Flumes and weirs

Flumes and weirs utilise the stage–discharge relationship described above but go a step further towards providing a continuous record of river discharge. If we think of stream discharge as consisting of a river velocity flowing through a cross-sectional area (as in the velocity profile method) then it is possible to isolate both of these terms separately. This is what flumes and weirs, or *stream gauging structures*, attempt to do.

The first part to isolate is the stream velocity. The way to do this is to slow a stream down (or, in some rare cases, speed a stream up) so that it flows with constant velocity through a known cross-sectional area. The critical point is that in designing a flume or weir the river flows at the same velocity (or at least a known velocity) through the gauging structure irrespective of how high the river level is. Although this seems counter-intuitive (rivers normally flow faster during flood events) it is achiev-

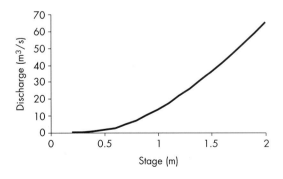

Figure 6.5 A rating curve for the river North Esk in Scotland based on stage (height) and discharge measurements from 1963–90.

able if there is an area prior to the gauging structure that slows the river down: a stilling pond.

One of the difficulties in maintaining gauging structures is that by slowing the river down in a stilling pond any sediment being carried by the water may be deposited, which with time will fill the stilling pond and lessen its usefulness. Because of this the stilling pond needs to be dredged regularly, particularly in a high energy environment such as mountain streams. To overcome this difficulty there is a design of trapezoidal flume that speeds the stream up rather than slows it down (see Figure 6.7). The stream is forced to go down a steep section immediately prior to the gauging structure. In this way any sediment is flushed out of the

weir, removing the need for regular dredging. This is really only possible for small streams as the power of large rivers at high velocities would place enormous strains on the gauging structure.

The second part of using a gauging structure is to isolate a cross-sectional area. To achieve this a rigid structure is imposed upon the stream so that it always flows through a known cross-sectional area. In this way a simple measure of stream height through the gauging structure will give the cross-sectional area. Stream height is normally derived through a stilling well, as described in Figure 6.6, except in this case there is a regular cross-sectional area.

Once the velocity and cross-sectional area are kept fixed the rating curve can be derived through a mixture of experiment and hydraulic theory. These relationships are normally power equations dependent on the shape of cross-sectional area used in the flume or weir.

The shape of cross-sectional area to be measured by stage height is an important consideration in the design of flumes and weirs. The shape of permanent structure that the river flows through is determined by the flow regime of the river and the requirements for the streamflow data. A common shape used is based on the V-structure (see Figure 6.8). The reason for this is that when river levels are low, a small

Figure 6.7 A trapezoidal flume. The stream passes through the flume and the water level at the base of the flume is recorded continuously.

Figure 6.8 A V-notch weir. The water level in the pond behind the weir is recorded continuously.

change in river flow will correspond to a significant change in stage (measured in the stilling well). This sensitivity to low flows makes data from this type of flume or weir particularly suitable for studying low flow hydrology. It is important that under high flow conditions the river does not overtop the flume or weir structure. The V shape is convenient for this also because as discharge increases the cross-sectional area flowed through increases in a non-linear fashion.

The difference between flumes and weirs

Although flumes and weirs perform the same function – measuring stream discharge in a continuous fashion – they are not exactly the same. In a weir the water is forced to drop over a structure (the weir) in the fashion of a small waterfall. In a flume the water passes through the structure without having a waterfall at the end.

Ultrasonic flow gauging

Recent technological developments have led to the introduction of a method of measuring stream discharge using the properties of sound wave propagation in water. The method actually measures water velocity, but where the stream bed cross-sectional area is known (and constant) the instrumentation can be left in place to provide a continuous measurement of river flow. There are two types of **ultrasonic flow gauges** that work in slightly different ways.

The first method measures the time taken for an ultrasonic wave emitted from a transmitter to reach a receiver on the other side of a river. The faster the water speed, the greater the deflection of the wave path and the longer it will take to cross the river. Sound travels at approximately 1,500 m/s in water (dependent on water purity and depth) so the instrumentation used in this type of flow gauging needs to be extremely precise and be able to measure in nanoseconds. This type of flow gauging can be installed as a permanent device but needs a width of river greater than 5 m and becomes unreliable with a high level of suspended solids.

The second method utilises the Doppler effect to measure the speed of particles being carried by the stream. At an extremely simple level this is a measurement of the wavelength of ultrasonic waves that bounce off suspended particles – the faster the particle the shorter the wavelength. This type of instrument works in small streams (less than 5 m width) and requires some suspended matter.

Hillslope runoff measurement

Overland flow

The amount of water flowing over the soil surface can be measured using collection troughs at the bottom of hillslopes or runoff plots. It is normal to use several troughs to characterise overland flow on a slope as it varies considerably in time and space. This spatial and temporal variation may be overcome with the use of a rainfall simulator.

Throughflow

Measurement of throughflow is fraught with difficulty. The only way to measure it is with throughflow troughs dug into the soil at the appropriate height. The problem with this is that in digging, the soil profile is disturbed and consequently the flow characteristics change. It is usual to insert troughs into a soil face that has been excavated and then fill the hole back up. This may still overestimate throughflow as the reconstituted soil in front of the troughs may encourage flow towards it as an area allowing rapid flow.

ESTIMATING STREAMFLOW

In the past 30 years probably the greatest effort in hydrological research has gone into creating numerical models to simulate streamflow. With time these have developed into models simulating all the processes in the hydrological cycle so that far more than just streamflow can be estimated. However, it is often streamflow that is seen as the end-product of a model, a reflection of the importance streamflow has as a hydrological parameter.

The easiest way of thinking about a hydrological model is to envisage streamflow as a series of numbers. Each number represents the volume of water that has flowed down the stream during a certain time period. A numerical model attempts to produce its own set of numbers, 'simulating' the flow of water down the river. There are many different ways of achieving this simulation, as will be discussed in the following section.

Numerical modelling strategies

Black box models

The simplest forms of numerical models simulate streamflow as a direct relationship between it and another measured variable. As an example a relationship can be derived between annual rainfall and annual runoff for a catchment (see Figure 6.9). The regression line drawn to correlate rainfall and runoff is a simulation model. If you know the annual rainfall for the catchment then you can simulate the annual runoff, using the regression relationship. This type of model is referred to as a black box model as it puts all the different hydrological processes that we know influence the way that water moves from rainfall to runoff into a single regression relationship. The simplicity of this type of model makes it widely applicable but its usefulness is restricted by the end-product from the model. In the example given, the regression model may be useful to estimate annual runoff in areas with the same geology and land use but it will not tell you anything about runoff at timescales less than one year or under different climatic and geomorphologic conditions.

Lumped conceptual models

Lumped conceptual models were the first attempt to reproduce the different hydrological processes within a catchment in a numerical form. Rainfall is added to the catchment and a water budget approach used to track the losses (e.g. evaporation) and movements of water (e.g. to and from soil water storage) within the catchment area. There are many examples in the literature of lumped conceptual models used to predict streamflows (e.g. Brandt et al., 1988).

The term 'lumped' is used because all of the processes operate at one spatial scale – that is, they are lumped together and there is no spatial discretisation. The scale chosen is often a catchment or sometimes sub-catchments.

The term 'conceptual' is used because the equations governing flow rates are often deemed to be conceptually similar to the physical processes operating. So, for instance, the storage of water in a canopy or the soil may be thought of as similar to storage within a bucket. As water enters the bucket it fills up until it overflows water at a rate equal to the entry rate. At the same time it is possible to have a 'hole' in the bucket that allows flow out at a rate dependent on the level of water within the bucket (faster with more water). This is analogous to soil water or canopy flow but is not a detailed description such as the Darcy–Richards approximation or the Rutter model. The rate of flow through the catchment, and hence the estimated streamflow, is

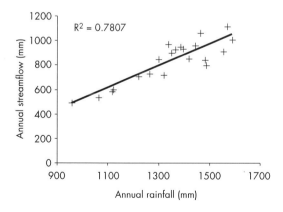

Figure 6.9 Annual rainfall vs. runoff data (1980–2000) for the Glendhu tussock catchment in the South Island of New Zealand.
Source: Data courtesy of Landcare Research NZ

controlled by a series of parameters that need to be calibrated for a given catchment. Calibration is normally carried out by comparing predicted flows to measured values and adjusting (or 'optimising') the parameters until the best fit is obtained. There is considerable debate on this technique as it may sometimes be possible to obtain a similar predicted hydrograph using a completely different set of optimised parameters. It is certainly true that the optimised parameters cannot be treated as having any physical meaning and should not be transferred to catchments other than those used for calibration.

Lumped conceptual models offer a method of formulating the hydrological cycle into a water budget model that allows simulation of streamflow while also being able to 'see' the individual processes operating. This is an advance beyond black box modelling, but because the processes are represented conceptually they are sometimes referred to as grey box models (i.e. you can see partially into them).

Physically based distributed models

The rapid advancements in computing power that have occurred since the 1970s mean that numerical modelling has become much easier. Freeze and Harlan (1969) were the first to formulate the idea of a numerical model that operates as a series of differential equations in a spatially distributed sense, an idea that prior to computers was unworkable. Their ideas (with some modifications from more recent research) were put into practice by several different organisations to make a physically based distributed hydrological model. Perhaps the best known of these is the Système Hydrologique Européen (SHE) model, built by a consortium of French, Danish and British organisations during the 1970s and early 1980s (Abbott *et al.*, 1986). A model such as the SHE uses many of the process estimation techniques described in earlier chapters (e.g. Darcy's law for subsurface flow, Rutter's model for canopy interception, snow-melt routines, etc.) in a water budgeting framework. Each of the equations or models used are solved for individual points within a catchment, using a grid pattern.

The principle behind this type of model is that it is totally transparent; all processes operating within a catchment are simulated as a series of physical equations at points distributed throughout the catchment. In theory this should mean that no calibration of the model is required and spatially distributed model output for any parameter can be obtained. In reality this is far from the case. There are numerous problems associated with using a physically based, distributed model, as outlined by Beven (1989), Grayson *et al.* (1992) and others. The principal problem is that the amount of data required to set the initial conditions and parameterise the model is vast. The idea of obtaining saturated hydraulic conductivity measurements for every grid point in a catchment is impossible, let alone all the other parameters required. The lack of data to run the model leads to spatial averaging of parameters. There are also concerns with the size of grid used in applications (sometimes up to 1 km²) and whether it is feasible to use the governing equations at this scale. These types of problems led Beven (1989) to query whether there really is such a thing as physically based distributed hydrological models or whether they are really just lumped conceptual models with fancier equations.

The concept of physically based distributed hydrological modelling is noble, but in reality the models have not produced the results that might have been expected. They are certainly unwieldy to use and have many simplifications that make the terminology doubtful. However they have been useful for gaining a greater understanding of our knowledge base in hydrological processes. The approach taken, with its lack of reliance on calibration, still offers the only way of investigating issues of land use change and ungauged catchments.

Hydrological modelling for specific needs

In many cases where streamflow is needed to be estimated, the use of a physically based model is akin to using the proverbial 'sledgehammer to crack a walnut'. With the continuing increase in computing power there are numerous tools available to the hydrologist to build their own computer model to simulate a particular situation of interest. These tools range from Geographic Information Systems (GIS) with attached dynamic modelling languages to object-oriented languages that can use icon-linked modelling approaches (e.g. McKim *et al.*, 1993). This perhaps offers a future role for hydrological modelling away from the large modelling packages such as SHE. In essence it allows the hydrologist to simulate streamflow based on a detailed knowledge of catchment processes of importance for the particular region of interest.

Physical or geomorphological estimation techniques

The geomorphological approach to river systems incorporates the idea that the river channel is in equilibrium with the flow regime. This suggests that measures of the channel (e.g. depth/width ratio, **wetted perimeter**, height to **bankfull discharge**) can be used to estimate the streamflows in both a historical and contemporary sense. Wharton (1995) provides a review of these different techniques. This is not a method that can be used to estimate the discharge in a river at one particular time, but it can be used to estimate parameters such as the mean annual **flood**. Important parameters to consider are the stream diameter, wetted perimeter and average depth. This is particularly for the area of channel that fills up during a small flooding event: so-called bankfull discharge.

It is possible to estimate the average velocity of a river stretch using a wave equation such as Manning's:

$$v = \frac{k.R^{2/3}.\sqrt{s}}{n}$$

where v is velocity (m/s); k is a constant depending on which units of measurement are being used (1 for SI units, 1.49 for Imperial); R is the **hydraulic radius** (m); s is the slope (m/m); and n is the Manning roughness coefficient. Hydraulic radius is the wetted perimeter of a river divided by the cross-

sectional area. In very wide channels this can be approximated as mean depth (Goudie *et al.*, 1994). The Manning roughness coefficient is estimated from knowledge of the channel characteristics (e.g. vegetation and bed characteristics) in a similar manner to Chezy's roughness coefficient in Table 6.3. Tables of Manning roughness coefficient can be found in Richards (1982), Chow *et al.* (1988), Goudie *et al.* (1994), and in other fluvial geomorphological texts.

FLOODS

The term *flood* is difficult to define except in the most general of terms. In a river it is normally considered to be an inundation caused by a period of abnormally large discharge (see Plate 6), but even this definition is fraught with inaccuracy. Flooding may occur from sources other than rivers (e.g. the sea and lakes), and 'abnormal' is difficult to pin down, particularly within a timeframe. Floods come to our attention through the amount of damage that they cause and for this reason they are often rated on a cost basis rather than on hydrological criteria. Hydrological and monetary assessment of flooding often differ markedly because the economic valuation is highly dependent on location. If the area of land inundated by a flooding river is in an expensive region with large infrastructure then the cost will be considerably higher than, say, for agricultural land. Two examples of large-scale floods during the 1990s illustrate this point. In 1998 floods in China caused an estimated US$20 billion of damage with over 15 million people being displaced and 3,000 lives lost (Smith, 2001). This flood was on a similar scale to one that occurred in the same region during 1954. A much larger flood in the Mississippi and Missouri rivers during 1993 resulted in a similar economic valuation of loss (US$15–20 billion) but only 48 lives were lost (USCE, 1996). The flood was the highest on record and has an average recurrence interval of between 100 and 500 years (USCE, 1996). The difference in lives lost and relative economic loss (for size of flood) is a reflection of the differing response to the flood in two economically contrasting countries.

As described in Chapter 2 for precipitation, flooding is another example where the *frequency–magnitude relationship* is important. Small flood events happen relatively frequently whereas the really large floods occur rarely but cause the most damage. The methods for interpreting river flows that may be used for flood assessment are discussed in Chapter 7. They provide some form of objective flood size assessment, but their value is highly dependent on the amount of data available.

Causes of floods

There are numerous reasons why a river will flood and they almost always relate back to the processes found within the hydrological cycle. Fundamentally rivers flood when there is too much rainfall for the river to cope with. The extent and size of the flood can often be related to other contributing factors that increase the effect of high rainfall. Some of these factors are described here but all relate back to concepts introduced in earlier chapters detailing the processes found within the hydrological cycle. Flooding provides an excellent example of the importance of scale, introduced in Chapter 1. Many of the factors discussed here have an influence at the small scale (e.g. hillslopes or small research catchments of less than 10 km^2) but not at the larger overall river catchment scale.

Other causes of floods are individual events like dam bursts, **jökulhlaups** (ice-dam bursts) or snowmelt (see pp. 63–66).

Antecedent soil moisture

The largest influence on size of flood is the wetness of the soil immediately prior to the rainfall or snowmelt occurring. As described on p. 53, the amount of infiltration into a soil and subsequent storm runoff are highly dependent on the degree of saturation in the soil. Almost all major flood events are heavily influenced by the amount of rainfall that has occurred prior to the actual flood causing rainfall.

Deforestation

The effects of trees on runoff has already been described, particularly with respect to water resources. There is also considerable evidence that a large vegetation cover, such as forest, decreases the severity of flooding that may be expected. There are several reasons for this. The first has already been described, in that trees provide an intercepting layer for rainfall and therefore slow down the rate at which the water reaches the surface. This will lessen the amount of overland flow as it may be absorbed by the soil. The second factor is that forests often have a high organic matter in the upper soil layers which, as any gardener will tell you, is able to absorb more water. Again this lessens the amount of overland flow, although it may increase the amount of throughflow. Finally, the infiltration rates under forest soils are often higher, leading again to less saturation excess overland flow.

The removal of forests from a catchment area will increase the propensity for a river to flood and also increase the severity of a flood event. Conversely the planting of forests on a catchment area will decrease the frequency and magnitude of flood events. Fahey and Jackson (1997) show that after conversion of native tussock grassland to exotic pine plantations a catchment in New Zealand showed a decrease in the mean flood peaks of 55–65 per cent. Although data of this type look alarming they are almost always taken from measurement at the small research catchment scale. At the larger scale the influence of deforestation is much harder to detect (see Chapter 9).

Urbanisation

Urban areas have a greater extent of impervious surfaces than in most natural landforms. Consequently the amount of infiltration excess (Hortonian) overland flow is high. In addition to this, urban areas are often designed to have a rapid drainage system, taking the overland flow away from its source. This network of gutters and drains frequently leads directly to a river drainage system, delivering more

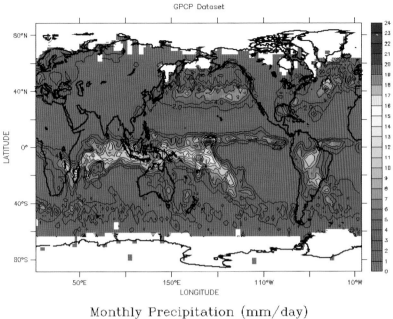

Monthly Precipitation (mm/day)

Plate 1 Satellite-derived global rainfall distribution in the month of January
Image from NOAA (www.ncdc.noaa.gov)

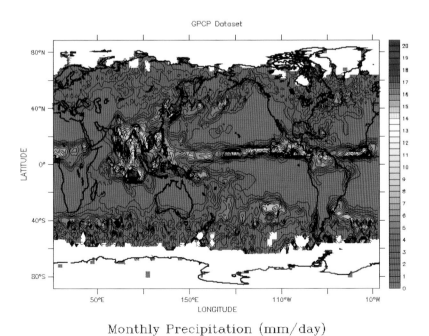

Monthly Precipitation (mm/day)

Plate 2 Satellite-derived global rainfall distribution in the month of July
Image from NOAA (www.ncdc.noaa.gov)

Plate 3 Water droplets condensing on the end of tussock leaves during fog

Plate 4 Cloud forming above a forest canopy immediately following rainfall. This cloud is formed by the evaporation of water intercepted by canopy leaves

Plate 5 Ice dam forming in a river in Canada
Photograph courtesy of Environment Canada

Plate 6 A river in flood. The excess water has spread across the flood plain outside the main river channel

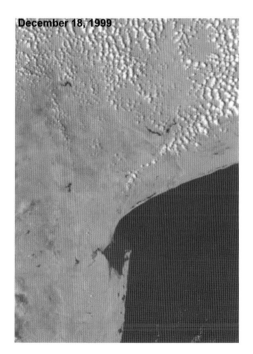

Plate 7 Satellite image of southern Mozambique prior to the flooding of 2000. Location can be compared from Figure 6.10
Image courtesy of G. Robert Brakenridge at the NASA-supported Dartmouth Flood Observatory

Plate 8 Satellite image of southern Mozambique following Cyclone Eline. The extensive flooding on the Incomáti and Limpopo (centre of image) can be seen clearly
Image courtesy of G. Robert Brakenridge at the NASA-supported Dartmouth Flood Observatory

Plate 9 The Nashua river during 1965, prior to water pollution remediation measures being taken
Photo courtesy of the Nashua River Watershed Association

Plate 10 The Nashua river during the 1990s, after remediation measures had been taken
Photo courtesy of the Nashua River Watershed Association

flood water in a faster time. Where extensive urban-isation of a catchment has occurred the flood frequency and magnitude has increased. Cherkauer (1975) shows a massive increase in flood magnitude for an urban catchment in Wisconsin, USA when compared to a similar rural catchment (see pp. 143–144). Urbanisation is another influence on flooding that is most noticeable at the small scale. This is mostly because the actual percentage area covered by impermeable urban areas in a larger river catchment is still very small in relation to the amount of permeable non-modified surfaces.

River channel alterations

Geomorphologists traditionally view a natural river channel as being in equilibrium with the river flowing within it. This does not mean that a natural river channel never floods, but rather that the channel has adjusted in shape in response to the normal discharge expected to flow through it. When the river channel is altered in some way it can have a detrimental effect on the flood characteristics for the river. In particular, **channelisation** using rigid structures can increase flood risk. Ironically, channeli-sation is often carried out to lessen flood risk in a particular area, but in actual fact all it does is pass the water on downstream at a faster rate than normal, increasing the flood risk further downstream. If there is a natural floodplain further downstream this may not be a problem, but if there is not, downstream riparian zones will be at greater risk.

Land drainage

It is common practice in many regions of the world to increase agricultural production through the drainage of 'swamp' areas. During the seventeenth and eighteenth centuries huge areas of the fenlands of East Anglia in England were drained and now are highly productive cereal and horticultural areas. The drainage of these regions provides for rapid removal of any surplus water, i.e. not needed by plants. Drained land will be drier than might be expected naturally, and therefore less storm runoff might be

assumed. This is true in small rainfall events but the rapid removal of water leads to flood peaks in the river drainage system where normally the water would have been slower to leave the land surface. Although the drainage of land leads to an overall drying out of the affected area it can also lead to increased flooding through rapid drainage. Again this is scale-related, as described further in Chapter 9.

Climate change

In recent years any flooding event has led to a clamour of calls to explain the event in terms of climatic change. This is not easy to do as climate is naturally so variable. What can be said though is that river channels slowly adjust to changes in flow regime which may in turn be influenced by changes in climate. Many studies have suggested that future climate change will involve greater extremes of weather (IPCC, 2001), including more high intensity rainfall events. This is likely to lead to an increase in flooding, particularly while a channel adjusts to the differing flow regime (if it is allowed to).

SUMMARY

The water flowing down a river is the end-product of precipitation after all the other hydrological processes have been in operation. The sub-processes of overland flow, throughflow and groundwater flow are well understood, although it is not easy to estimate their relative importance for a particular site, particularly during a storm event. The measurement of river flow is relatively straightforward and presents the fewest difficulties in terms of sampling error, although there are limitations, particularly during periods of high flow and floods.

Case study

MOZAMBIQUE FLOODS OF 2000

During the early months of 2000 world news was dominated by the catastrophic flooding that occurred in southern Africa and Mozambique in particular. The most poignant image from this time was the rescuing of a young mother, Sophia Pedro, with her baby Rosita, born up a tree while they sought refuge from the flood waters. The international media coverage of the devastating flood damage and the rescue operation that followed has ensured that this flood will be remembered for a long time to come. It has given people the world over a reminder that flooding is a hydrological hazard capable of spreading devastation on a huge scale.

The floods of Mozambique were caused by four storms in succession from January through to March 2000. The first three months of the year are the rainy season (or monsoon) for south-eastern Africa and it is usual for flooding to occur, although not to the scale witnessed in 2000. The monsoon started early in southern Mozambique; the rainfall in Maputo was 70 per cent above normal for October–November 1999. This meant that any heavy rainfall later in the rainy season would be more likely to cause a flood.

The first flood occurred during January 2000 when the Incomáti and Maputo rivers (see Figure 6.10) both burst their banks causing widespread disruption. The second flood occurred in early February, as the waters started to recede, except that now Cyclone Connie brought record rainfall to southern Mozambique and northern South Africa. The Limpopo river was as high as ever recorded (previous high was in 1977) and major communication lines were cut. The third flood, 21 February until the end of February, occurred when Cyclone Eline moved inland giving record rainfall in Zimbabwe and northern South Africa, causing record-breaking floods. The Limpopo was 3 metres higher than any recorded flood and for the first time in recorded history the Limpopo and Incomáti rivers joined together in a huge inundation. The extent of the flooding can be seen in the satellite images (see Plates 7 and 8). The fourth flood was similar in size to the second and occurred following Cyclone Glória in early March (Christie and Hanlon, 2001).

There is no doubt that the Mozambique floods were large and catastrophic. How large they are, in terms of return periods or average recurrence intervals (see Chapter 7) is difficult to assess. The major difficulty is to do with paucity of streamflow records and problems with measuring flows during flood events. On the Incomáti river the flow records go back to 1937, and this was the largest flood recorded. For the Limpopo there is some data back to the 1890s, and again this was the largest recorded flow event. On the Maputo river to the south the flood levels were slightly lower than a 1984 event. The difficulties in measuring riverflow during large flood events is well illustrated by the failure of many gauging stations to function properly, either through complete inundation or being washed away. Christie and Hanlon (2001) quote an estimate of the flood on the Limpopo having a 100-year average recurrence interval, although this is difficult to verify as most gauges failed. Smithers et al. (2001) quote an unpublished report by Van Bladeren and Van der Spuy (2000) suggesting that upstream tributaries of the Incomáti river exceeded the 100-year return period. Smithers et al. (2001) provide an analysis of the 1–7 day rainfall for the Sabie catchment (a tributary of the Incomáti) which shows that in places the 200-year return period was exceeded. (NB This is an analysis of rainfall records not riverflow.)

The reasons for the flooding were simple, as

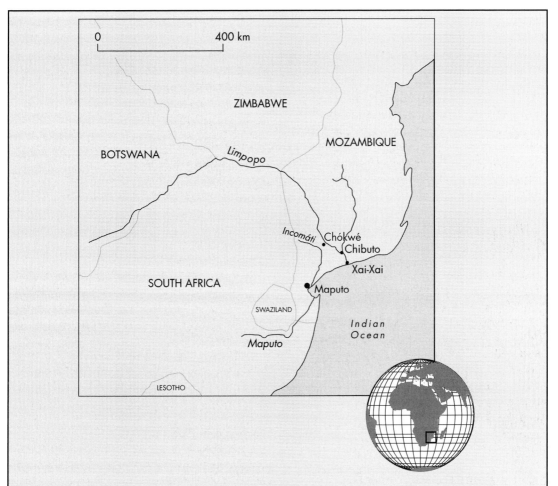

Figure 6.10 Location of the Incomáti, Limpopo and Maputo rivers in southern Africa.

they are in most cases: there was too much rainfall for the river systems to cope with the resultant stormflow. The river catchments were extremely wet (i.e. high antecedent soil moisture values) prior to the extreme rainfall, due to a prolonged and wet monsoon. One possible explanation for the severity of the rainfall is linked in with the ENSO (El Niño: Southern Oscillation) ocean-weather patterns in the Pacific. Christie and Hanlon (2001) present evidence that during a La Niña event (extreme cold temperatures in the western Pacific Ocean) it is common to see higher rainfall totals in Mozambique. However, this is not a strong relationship and certainly could not be used to make predictions. Figure 6.11 shows the monsoon rainfall at Maputo (averaged over two rainy seasons) and associated La Niña events. There may be some link here but it is not immediately obvious, particularly when you consider 1965–66 which had high rainfall despite it being an El Niño event (often associated with drought in southern Africa).

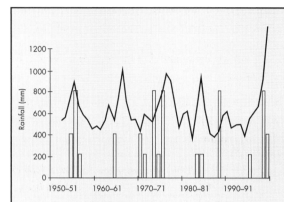

Figure 6.11 Rainfall totals during the rainy season (smoothed with a two-year average) at Maputo airport, with vertical bars indicating the strength of La Niña events (on a scale of three: strong, medium, weak).
Sources: Rainfall data from Christie and Hanlon (2001); La Niña strength from NOAA

floods in the headwaters of rivers draining into Mozambique. Flood warnings were issued by Zimbabwe and South Africa but the poor state of communications in Mozambique (exacerbated by the previous floods cutting communication lines) meant that they were not available to warn people on the ground. In all 700 people died as a result of the floods and 45,000 people were displaced. It is estimated that it will cost US$450 million to repair damage to the infrastructure in Mozambique (Christie and Hanlon, 2001). This is not the total cost of the flood, which is far higher when loss of income and loss of private property is included. These costs will never be fully known as in many lesser-developed countries the costs are borne by individuals without any form of insurance cover.

In many ways there are no new lessons to learn from the Mozambique floods of 2000. It is well known that adequate warning systems are needed (but expensive) and that people should be restricted from living in flood-prone areas; but this is difficult to achieve in a poor country such as Mozambique. The cause of the flood was a huge amount of rainfall and the severity was influenced by the antecedent wetness of the ground due to a very wet monsoon.

What was unusual about the 2000 floods was that the tropical Eline cyclone (called typhoons or hurricanes elsewhere) moved inland, taking extremely high rainfall to Zimbabwe and northern South Africa. This is not normal behaviour for this type of storm and in so doing it created large

ESSAY QUESTIONS

1 Outline the different theories of storm runoff generation and assess their relative importance for a small catchment near you.

2 Compare and contrast the different methods for measuring instantaneous streamflow.

3 Describe the different methods available for measuring long-term streamflow (e.g. hourly for several years).

4 Explain the difference in approach between black box and physically based distributed hydrological models.

5 Discuss the importance of spatial scale in assessing causes of flooding.

FURTHER READING

Anderson M.G. and Burt, T.P. (eds) (1990) *Process studies in hillslope hydrology*. Wiley, Chichester.
A slightly more modern update on Kirkby (1978).

Beven, K.J. (2001) *Rainfall-runoff modelling: the primer*. Wiley, Chichester.
An introduction to modelling in hydrology.

Kirkby, M.J. (ed.) (1978) *Hillslope hydrology*. J. Wiley & Sons, Chichester.
A classic text on hillslope processes, particularly runoff.

Parsons, A.J. and Abrahams, A.D. (1992) *Overland flow: hydraulics and erosion mechanics*. UCL Press, London.
An advanced edited book; good detail on arid regions.

Smith, K. and Ward, R.C. (1998) *Floods. Physical processes and human impacts*. Wiley, Chichester.
A text on flooding.

Wohl, E.E. (2000) *Inland flood hazards: human, aquatic and riparian communities*. Cambridge University Press, Cambridge.
A text on flooding with many case studies.

7

STREAMFLOW ANALYSIS

LEARNING OBJECTIVES

When you have finished reading this chapter you should have:

- An understanding of what different hydrological techniques are used for.
- A knowledge of hydrograph analysis (including the unit hydrograph).
- A knowledge of how to derive and interpret flow duration curves.
- A knowledge of how to carry out frequency analysis, particularly for floods.

One of the most important tasks in hydrology is to analyse streamflow data. These data are continuous records of discharge, frequently measured in permanent structures such as flumes and weirs (see Chapter 6). Analysis of these data provides us with three important features:

- description of a flow regime
- potential for comparison between rivers, and
- prediction of possible future river flows.

There are well-established techniques available to achieve these, although they are not universally applied in the same manner. This chapter sets out three important methods of analysing streamflow: hydrograph analysis, flow duration curves and frequency analysis. These three techniques are explained with reference to worked examples, all drawn from the same data set. The use of data from within the same study area is important for comparison between the techniques.

HYDROGRAPH ANALYSIS

A hydrograph is a continuous record of stream or river discharge (see Figure 6.1). It is a basic working unit for a hydrologist to understand and interpret. The shape of a hydrograph is a response from a particular catchment to a series of unique conditions, ranging from the underlying geology and catchment shape to the antecedent wetness and storm duration.

The temporal and spatial variations in these underlying conditions make it highly unlikely that two hydrographs will ever be the same. Although there is great variation in the shape of a hydrograph there are common characteristics of a storm hydrograph that can be recognised. These have been described in Chapter 6 where terms such as 'rising limb', 'falling limb', 'recession limb' and 'baseflow' are explained.

Hydrograph separation

The separation of a hydrograph into baseflow and stormflow is a common task, although never particularly easy. The idea of **hydrograph separation** is to distinguish between stormflow and baseflow so that the amount of water resulting from a storm can be calculated. Sometimes further assumptions are made concerning where the water in each component has come from (i.e. groundwater and overland flow) but, as explained in the previous chapter, this is controversial.

The simplest form of hydrograph separation is to draw a straight, level line from the point where the hydrograph starts rising until the stream discharge reaches the same level again (dashed line in Figure 7.1). However, this is frequently problematic as the stream may not return to its pre-storm level before another storm arrives. Equally the storm may recharge the baseflow enough so that the level is raised after the storm (as shown in Figure 7.1).

To overcome the problem of a level baseflow separation a point has to be chosen on the receding limb where it is decided that the discharge has returned to baseflow. Exactly where this point will be is not an easy to determine. By convention the point is taken where the recession limb fits an exponential curve. This can be detected by plotting the natural log (ln) of discharge (Q) and noting where this line becomes straight. The line drawn between the start and 'end' of a storm may be straight (dotted line) or curved (thin solid line) depending on the preference of hydrologist.

In very large catchments there is a formula that can be applied to derive the time where stormflow ends. This is the fixed time method which gives the time from peak flow to the end of stormflow (τ):

$$\tau = D^n$$

where D is the drainage area and n is a recession constant. When D is in square miles and τ in days, the value of n has been found to be approximately 0.2.

The problem with hydrograph separation is that the technique is highly subjective. There is no physical reasoning why the 'end' of a storm should be when the recession limb fits an exponential curve; it is pure convention. Equally the shape of the curve between start and 'end' has no physical reasoning. It does not address the debate covered in Chapter 6: where does the stormflow water come from? What hydrograph separation does offer is a means of separating stormflow from baseflow, something that is needed for the use of the unit hydrograph (see pp. 91–94), and may be useful for hydrological interpretation and description.

The unit hydrograph

The idea of a **unit hydrograph** was first put forward by Sherman, an American engineer working in the 1920s and 1930s. The idea behind the unit hydrograph is simple enough, although it is a somewhat tedious exercise to derive one for a catchment. The fundamental concept of the unit hydrograph is

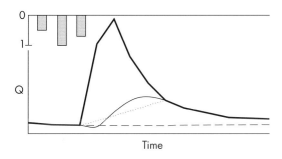

Figure 7.1 Hydrograph separation techniques. See text for explanation.

that the shape of a storm hydrograph is determined by the physical characteristics of the catchment. The majority of those physical characteristics are static in time, therefore if you can find an average hydrograph for a particular storm size then you can use that to predict other storm events. In short: two identical rainfall events that fall on a catchment with exactly the same antecedent conditions should produce identical hydrographs.

With the unit hydrograph a hydrologist is trying to predict a future storm hydrograph that will result from a particular storm. This is particularly useful as it gives more than just the peak runoff volume and includes the temporal variation in discharge.

Sherman (1932) defines a unit hydrograph as 'the hydrograph of surface runoff resulting from **effective rainfall** falling in a unit of time such as 1 hour or 1 day'. The term effective rainfall is taken to be that rainfall that contributes to the storm hydrograph. This is often assumed to be the rainfall that does not infiltrate the soil and moves into the stream as overland flow. This is infiltration excess, or Hortonian overland flow, Sherman's ideas fitting in well with those of Horton. Sherman assumed that the 'surface runoff is produced uniformly in space and time over the total catchment area'.

Deriving the unit hydrograph: step 1

Take historical rainfall and streamflow records for a catchment and separate out a selection of typical single-peaked storm hydrographs. It is important that they are separate storms as the method assumes that one runoff event does not affect another. For each of these storm events separate the baseflow from the stormflow; that is, hydrograph separation (see p. 91). This will give you a series of storm hydrographs (without the baseflow component) for a corresponding rainfall event.

Deriving the unit hydrograph: step 2

Take a single storm hydrograph and find out the total volume of water that contributed to the storm. This can be done either by measuring the area under the stormflow hydrograph or as an integral of the curve. If you then divide the total volume in the storm by the catchment area you have the runoff as a water equivalent depth. If this is assumed to have occurred uniformly over space and time within the catchment then you can assume it is equal to the effective rainfall. This is an important assumption of the method: that the effective rainfall is equal to the water equivalent depth of storm runoff. It is also assumed that the effective rainfall all occurred during the height of the storm (i.e. during the period of highest rainfall intensity). That period of high rainfall intensity becomes the time for the unit hydrograph.

Deriving the unit hydrograph: step 3

The unit hydrograph is the stormflow that results from one unit of effective rainfall. To derive this you need to divide the values of stormflow (i.e. each value on the storm hydrograph) by the amount of effective rainfall (from step 2) to give the unit hydrograph. This is the discharge per millimetre of effective rainfall during the time unit.

Deriving the unit hydrograph: step 4

Repeat steps 2 and 3 for all of the typical hydrographs. Then create an average unit hydrograph by merging the curves together. This is achieved by averaging the value of stormflow for each unit of time for every derived unit hydrograph. It is also possible to derive different unit hydrographs for different rain durations and intensities, but this is not covered in this text (see Chow *et al.*, 1998, or Shaw, 1994, for more details).

Using the unit hydrograph

The unit hydrograph obtained from the steps described here theoretically gives you the runoff that can be expected per mm of effective rainfall in one hour. In order to use the unit hydrograph for predicting a storm it is necessary to estimate the 'effective rainfall' that will result from the storm

Worked example of the unit hydrograph

The Tanllwyth is a small (0.98 km²) headwater tributary of the Severn river in mid-Wales. The catchment is monitored by the Centre for Ecology and Hydrology (formerly the Institute of Hydrology) as part of the Plynlimon catchment experiment. For this example a storm was chosen from July 1982 as it is a simple single-peaked hydrograph (see Figure 7.2).

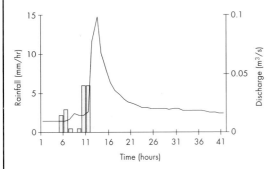

Figure 7.2 A simple storm hydrograph (July 1982) from the Tanllwyth catchment.

Baseflow separation was carried out by using a straight-line method. The right-hand end of the straight line (shown as a dashed line in Figure 7.3) is where the receding limb of the hydrograph became exponential.

All of the flow above the dashed line in Figure 7.3 was then divided by the effective rainfall to derive the unit hydrograph in Figure 7.4 below.

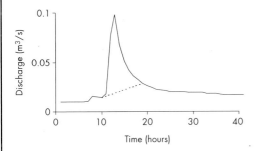

Figure 7.3 Baseflow separation. The dashed line meets the solid line at the point where the receding limb becomes exponential.

Figure 7.4 The unit hydrograph for the Tanllwyth catchment.

NB The hydrograph appears more spread out because of the scale of drawing.

To apply the unit hydrograph to a small storm, hydrographs were added together for each amount of effective rainfall. The resultant total hydrograph is shown as a dark black line in Figure 7.5. The discharge values in the simulated hydrograph are much larger than those in the original storm hydrograph despite what appears to be a smaller storm. This is because the simulated hydrograph is working on effective rainfall rather than actual rainfall. Effective rainfall is the rain that does not infiltrate and is theoretically available for storm runoff. A low effective rainfall value may represent a high actual rainfall value.

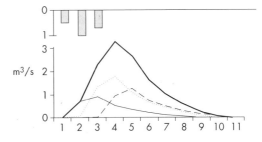

Figure 7.5 Applying the unit hydrograph to a small storm (effective rainfall shown on the separate scale above). Three different lines represent the flow from each of the rainfall bars (thin solid first, then dotted, then dashed). The solid black line is the total discharge (i.e. the sum of the three lines).

rainfall. This is not an easy task and is one of the main hurdles in using the method. In deriving the unit hydrograph the assumption has been made that 'effective rainfall' is the rainfall which does not infiltrate but is routed to the stream as overland flow (Hortonian). The same assumption has to be made when utilising the unit hydrograph. To do this it is necessary to have some indication of the infiltration characteristics for the catchment concerned, and also of the antecedent soil moisture conditions. The former can be achieved through field experimentation and the latter through the use of an antecedent precipitation index (API). Engineering textbooks give examples of how to use the API to derive effective rainfall. The idea is that antecedent soil moisture is controlled by how long ago rain has fallen and how large that event was. The wetter a catchment is prior to a storm, the more effective rainfall will be produced.

Once the effective rainfall has been established it is a relatively simple task to add the resultant unit hydrographs together to form the resultant storm hydrograph. The worked example shows how this procedure is carried out.

Limitations of the unit hydrograph

The unit hydrograph has several assumptions that at first appearance would seem to make it inapplicable in many situations. The assumptions can be summarised as:

- The runoff that makes up stormflow is derived from infiltration excess (Hortonian) overland flow. As described in Chapter 6, this is not a reasonable assumption to make in many areas of the world.
- That the surface runoff occurs uniformly over the catchment because the rainfall is uniform over the catchment. Another assumption that is difficult to justify.
- The relationship between effective rainfall and surface runoff does not vary with time (i.e. the hydrograph shape remains the same between the data period of derivation and prediction).

This would assume no land use change within the catchment, as this could well affect the storm hydrograph shape.

Given the assumptions listed above it would seem extremely foolhardy to use the unit hydrograph as a predictive tool. However, the unit hydrograph has been used successfully for many years in numerous different hydrological situations. It is a very simple method of deriving a storm hydrograph from a relatively small amount of data. The fact that it does work (i.e. produce meaningful predictions of storm hydrographs), despite being theoretically flawed, would seem to raise questions about our understanding of hydrological processes. The answer to why it works may well lie in the way that it is applied, especially the use of effective rainfall. This is a nebulous concept that is difficult to describe from field measurements. It is possible that in moving from actual to effective rainfall there is a blurring of processes that discounts some of the assumptions listed above. The unit hydrograph is a black box model of stormflow and as such hides many different processes within. The simple concept that the hydrograph shape is a reflection of the static characteristics and all the dynamic processes going on in a catchment makes it highly applicable but less able to be explained in terms of hydrological theory.

The synthetic unit hydrograph

The **synthetic unit hydrograph** is an attempt to derive the unit hydrograph from measurable catchment characteristics rather than from flow data. This is highly desirable as it would give the opportunity to predict stormflows when having no historical streamflow data; a common predicament around the world. The Institute of Hydrology in the UK carried out an extensive study into producing synthetic unit hydrographs for catchments, based on factors such as the catchment size, degree of urbanisation, etc. (NERC, 1975). They produced a series of multiple regression equations to predict peak runoff amount, time to peak flow, and the time to the end of the recession limb based on the measurable characteristics. Although this has been carried out relatively successfully it is only applicable to the UK as that is where the derivative data was from. In another climatic area the hydrological response is likely to be different for a similar catchment. The UK is a relatively homogeneous climatic area with a dense network of river flow gauging, which allowed the study to be carried out. In areas of the world with great heterogeneity in climate and sparse river monitoring it would be extremely difficult to use this approach.

FLOW DURATION CURVES

An understanding of how much water is flowing down a river is fundamental to hydrology. Of particular interest for both flood and low flow hydrology is the question of how representative a certain flow is. This can be addressed by looking at the frequency of daily flows and some statistics that can be derived from the frequency analysis. The culmination of the frequency analysis is a **flow duration curve** which is described below.

Flow duration curves are concerned with the amount of time a certain flow is exceeded. The data most commonly used are daily mean flows: the average flow for each day (note well that this is not the same as a mean daily flow, which is the average of a series of daily flows). To derive a flow duration curve the daily mean flow data are required for a long period of time, in excess of five years. A worked example is provided here, using 26 years of data for the upper reaches of the river Wye in mid-Wales, UK (see pp. 96–97).

Flow duration curve: step 1

A table is derived that has the frequency, cumulative frequency (frequency divided by the total number of observations) and percentage cumulative frequency. The percentage cumulative frequency is assumed to equal the percentage of time that the flow is exceeded. While carrying out the frequency analysis

it is important that a small class (or bin) interval is used; too large an interval and information will be lost from the flow duration curve. The method for choosing the best class interval is essentially through trial and error. As a general rule you should aim not to have more than around 10 per cent of your values within a single class interval. If you have more than this you start to lose precision in plotting. As shown in the worked example, it is not essential that the same interval is used throughout.

Flow duration curve: step 2

The actual flow duration curve is created by plotting the percentage cumulative frequency on the x-axis against the mid-point of the class interval on the y-axis. Where two flow duration curves are presented on the same axes they need to be standardised for direct comparison. To do this the values on the y-axis (mid-point of class interval) are divided by the average flow for the record length. This makes the y-axis a percentage of the average flow (see Figure 7.6).

The presentation of a flow duration curve may be improved by either plotting on a special type of graph paper or transforming the data. The type of graph paper often used has the x-axis transformed in the form of a known distribution such as the Gumbel

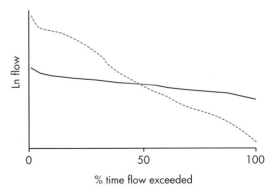

Figure 7.6 Two contrasting flow duration curves. The dotted line has a high variability in flow (similar to a small upland catchment) compared to the solid line (similar to a catchment with a high baseflow).

Worked example of flow duration curve

The river Wye has its headwaters in mid-Wales and flows into the Severn at the head of the Severn estuary. In its upper reaches it is part of the Plynlimon hydrological experiment run by the Institute of Hydrology (now the Centre for Ecology and Hydrology) from the early 1970s. At Plynlimon the Wye is a small (10.5 km²) grassland catchment with an underlying geology of relatively impermeable Ordovician shale. The data used to derive a flow duration curve here are from the upper Wye for a period from 1971 until 1995, consisting of 9,437 values of daily mean flow in cumecs.

The frequency analysis for the Wye is shown in Table 7.1.

The flow duration curve (see Figure 7.7) is derived by plotting the percentage cumulative frequency (x-axis) against the mid-point of the daily mean flow class intervals (y-axis). When this is plotted it forms the exponential shape that is normal for this type of catchment. In order to see more detail on the curve the flow values can be logged (natural log). This is shown in Figure 7.8.

The flow statistics Q_{95}, Q_{50} and Q_{10} can either be read from the graph (see Figure 7.9) or interpolated from the original frequency table (remembering to use the mid-points of the class interval).

A summary of the flow statistics for the upper Wye are shown in Table 7.2.

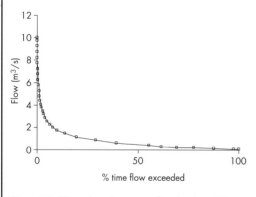

Figure 7.7 Flow duration curve for the river Wye (1971–95 data).

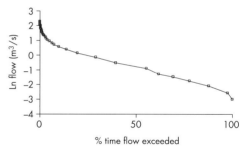

Figure 7.8 Flow duration curve for the river Wye (1971–95 data) with the flow data shown on a natural log scale.

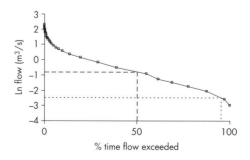

Figure 7.9 Q_{95} (short dashes) and Q_{50} (long dashes) shown on the flow duration curve.

Table 7.1 Values from the frequency analysis of daily mean flow on the upper Wye catchment. These values form the basis of the flow duration curve in Figure 7.7

Daily mean flow (m³/s)	Frequency	Relative frequency (%)	Cumulative frequency (%)
0–0.5	250	2.65	100.00
0.5–0.1	923	9.78	97.35
0.1–0.15	927	9.82	87.57
0.15–0.2	814	8.63	77.75
0.2–0.25	708	7.50	69.12
0.25–0.3	589	6.24	61.62
0.3–0.4	881	9.34	55.38
0.4–0.5	641	6.79	46.04
0.5–0.7	958	10.15	39.25
0.7–1.0	896	9.49	29.10
1.0–1.3	553	5.86	19.60
1.3–1.6	357	3.78	13.74
1.6–1.9	222	2.35	9.96
1.9–2.1	117	1.24	7.61
2.1–2.4	127	1.35	6.37
2.4–2.7	103	1.09	5.02
2.7–3.0	71	0.75	3.93
3.0–3.3	47	0.50	3.18
3.3–3.6	42	0.45	2.68
3.6–3.9	34	0.36	2.24
3.9–4.2	33	0.35	1.88
4.2–4.5	28	0.30	1.53
4.5–5.0	28	0.30	1.23
5.0–5.5	23	0.24	0.93
5.5–6.0	23	0.24	0.69
6.0–6.5	14	0.15	0.45
6.5–7.0	7	0.07	0.30
7.0–7.5	7	0.07	0.22
7.5–8.0	3	0.03	0.15
8.0–8.5	4	0.04	0.12
8.5–9.0	2	0.02	0.07
9.0–9.5	1	0.01	0.05
9.5–10.0	3	0.03	0.04
> 10.0	1	0.01	0.01
Total	9,437	100	

Table 7.2 Summary flow statistics derived from the flow duration curve for the Wye catchment

Flow statistic	Ln flow (m³/s)	Flow (m³/s)
Q_{95}	–2.48	0.084
Q_{50}	–0.78	0.458
Q_{10}	0.56	1.75

or Log Pearson. A natural log transformation of the flow values (y-axis) achieves a similar effect, although this is not necessarily standard practice.

Interpreting a flow duration curve

The shape of a flow duration curve can tell a lot about the hydrological regime of a catchment. In Figure 7.6 two flow duration curves of contrasting shape are shown. With the dotted line there is a large difference between the highest and lowest flow values, whereas for the solid line there is far less variation. This tells us that the catchment shown by the solid line never has particularly low flows or particularly high flows. This type of hydrological response is found in limestone or chalk catchments where there is a high baseflow in the summer (groundwater derived) and high infiltration rates during storm events. In contrast the catchment shown with a solid line has far more variation. During dry periods it has a very low flow, but responds to rainfall events with a high flow. This is characteristic of impermeable upland catchments or streamflow in dryland areas.

Statistics derived from a flow duration curve

The interpretation of flow duration curve shape discussed so far is essentially subjective. In order to introduce some objectivity there are statistics derived from the curve; the three most important ones are:

- The flow value that is exceeded 95 per cent of the time (Q_{95}). A useful statistic for low flow analysis.
- The flow value that is exceeded 50 per cent of the time (Q_{50}). This is the median flow value.
- The flow value that is exceeded 10 per cent of the time (Q_{10}). A useful statistic for analysis of high flows and flooding.

FREQUENCY ANALYSIS

The analysis of how often an event is likely to occur is an important concept in hydrology. It is the application of statistical theory into an area that affects many people's lives, whether it be through flooding or low flows and drought. Both of these are considered here, although because they use similar techniques the main emphasis is on **flood frequency analysis**. The technique is a statistical examination of the frequency–magnitude relationship discussed in Chapters 2 and 6. It is an attempt to place a probability on the likelihood of a certain event occurring. Predominantly it is concerned with the low-frequency, high-magnitude events (e.g. a large flood or a very low river flow).

It is important to differentiate between the uses of flow duration curves and frequency analysis. Flow duration curves tell us the percentage of time that a flow is above or below a certain level. This is average data and describes the overall flow regime. Flood frequency analysis is concerned only with peak flows: the probability of a certain flood recurring. Conversely, **low flow frequency analysis** is concerned purely with the lowest flows and the probability of them recurring.

Flood frequency analysis

Flood frequency analysis is probably the most important hydrological technique. The concept of a '100-year flood', or a 50-year recurrence interval, is well established in most people's perceptions of hydrology, although there are many misunderstandings in interpretation.

Flood frequency analysis is concerned with peak flows. There are two different ways that a peak flow can be defined:

- the single maximum peak within a year of record giving an **annual maximum series**; or
- any flow above a certain threshold value, giving a **partial duration series**.

Figure 7.10 shows the difference between these two

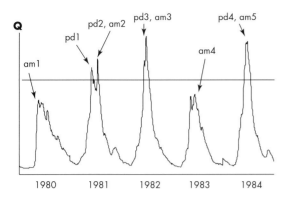

Figure 7.10 Daily flow record for the Adams river (British Columbia, Canada) during five years in the 1980s. Annual maximum series are denoted by 'am', partial duration series above the threshold line by 'pd'. NB In this record there are five annual maximum data points and only four partial duration points, including two from within 1981.
Source: Data courtesy of Environment Canada

data series. There are arguments for and against the use of either data series in flood frequency analysis. Annual maximum may miss a large storm event where it occurs more than once during a year (as in the 1981 case in Figure 7.10), but it does provide a continuous series of data that are relatively easy to process. The setting of a threshold storm (the horizontal line in Figure 7.10) is critical in analysis of the partial duration series, something that requires considerable experience to get right. The statistics used to analyse the two series are different due to the differing nature of the data. The most common analysis is on annual maximum series, the simplest form, which is described here. If the data series is longer than ten years then the annual maxima can be used; for very short periods of record the partial duration series can be used.

The first step in carrying out flood frequency analysis is to obtain the data series (in this case annual maxima). The annual maximum series should be for as long as the data record allows. The greater the length of record the more certainty can be attached to the prediction of average recurrence interval. Many hydrological database software pack-

ages (e.g. HYDSYS) will give annual maxima data automatically, but some forethought is required on what annual period is to be used. There is an assumption made in flood frequency analysis that the peak flows are independent of each other (i.e. they are not part of the same storm). If a calendar year is chosen for a humid temperate environment in the northern hemisphere, or a tropical region, it is possible that the maximum river flow will occur in the transition between years (i.e. December/January). It is possible for a storm to last over the 31 December/1 January period and the same storm to be the maximum flow value for both years. If the flow regime is dominated by snow-melt then it is important to avoid splitting the hydrological year at times of high melt (e.g. spring and early summer). To avoid this it is necessary to choose your hydrological year as changing during the period of lowest flow. This may take some initial investigation of the data.

All flood frequency analysis is concerned with the analysis of frequency histograms and probability distributions. Consequently the first data analysis step should be to draw a frequency histogram. It is often useful to convert the frequency into a relative frequency (divide the number of readings in each class interval by the total number of readings in the data series).

The worked example given is for a data set on the river Wye in mid-Wales (see pp. 102–103). On looking at the histogram of the Wye data set (Figure 7.11) the first obvious point to note is that it is not normally distributed (i.e. it is not a classic bell-shaped curve). It is important to grasp the significance of the non-normal distribution for two reasons:

1 Common statistical techniques that require normally distributed data (e.g. t-tests etc.) cannot be applied in flood frequency analysis.
2 It shows what you might expect: small events are more common than large floods, but that very large flood events do occur.

If you were to assume that the data series is infinitely

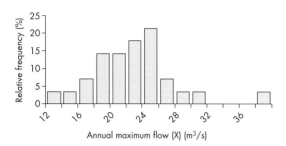

Figure 7.11 Frequency distribution of the Wye annual maximum series.

large in number and the class intervals were made extremely small, then a smooth curve can be drawn through the histogram. This is the *probability density function* which represents the smoothed version of your frequency histogram.

In flood frequency analysis there are three interrelated terms of interest. These terms are inter-related mathematically, as described below.

1 The probability of excedence: $P(X)$. This is the probability that a flow (Q) is greater than, or equal to a value X. The probability is normally expressed as a unitary percentage (i.e. on a scale between 0 and 1).
2 The relative frequency: $F(X)$. This is the probability of the flow (Q) being less than a value X. This is also expressed as a unitary percentage.
3 The average recurrence interval: $T(X)$. This is sometimes referred to as the return period, although this is misleading. $T(X)$ is a statistical term meaning the chance of excedence once every T years over a long record. This should not be interpreted as meaning that is exactly how many years are likely between certain size floods.

$$P(X) = 1 - F(X)$$

$$T(X) = \frac{1}{P(X)} = \frac{1}{1 - F(X)}$$

It is possible to read the values of $F(X)$ from a cumulative probability curve; this provides the simplest method of carrying out flood frequency analyses. One difficulty with using this method is that you must choose the class intervals for the histogram carefully so that the probability density function is an accurate representation of the data. Too large an interval and the distribution may be shaped incorrectly, too small and holes in the distribution will appear.

One way of avoiding the difficulties of choosing the best class interval is to use a rank order distribution. This is often referred to as a plotting position formula.

The Weibull formula

The first step in the method is to rank your annual maximum series data from low to high. In doing this you are assuming that each data point (i.e. the maximum flood event for a particular year) is independent of any others. This means that the year that the flood occurred in becomes irrelevant.

Taking the rank value, the next step is to calculate the $F(X)$ term using the formula given here. In this case r refers to the rank of an individual flood event (X) within the data series and N is the total number of data points (i.e. the number of years of record):

$$F(X) = \frac{r}{N+1}$$

In applying this formula there are two important points to note:

1 The value of $F(X)$ can never reach 1 (i.e. it is asymptotic towards the value 1).
2 If you rank your data from high to low (i.e. the other way around) then you will be calculating the $P(X)$ value rather than $F(X)$. This is easily rectified by using the formula linking the two.

The worked example on pp. 102–103 gives the $F(X)$, $P(X)$ and $T(X)$ for a small catchment in mid-Wales (Table 7.3).

The Weibull formula is simple to use and effective but is not always the best description of an annual maximum series data. Some users suggest that a better fit to the data is provided by the Gringorten formula:

$$F(X) = \frac{r - 0.44}{N + 0.12}$$

As illustrated in the worked example, the difference between these two formulae is not great and often the use of either one is down to personal preference.

Extrapolating beyond your data set

The probabilities derived from the Weibull and Gringorten formulae give a good description of the flood frequency within the measured stream record but do not provide enough data when you need to extrapolate beyond a known time series. This is a common hydrological problem: we need to make an estimate on the size of a flood within an average recurrence interval of 50 years but only have 25 years of streamflow record. In order to do this you need to fit a distribution to your data. There are several different ways of doing this, the method described here uses the method of moments based on the Gumbel distribution. Other distributions that are used by hydrologists include the Log-Pearson Type III and log normal. The choice of distribution is often based on personal preference and regional bias (i.e. the distribution that seems to fit flow regimes for a particular region).

Method of moments

If you assume that the data fits a Gumbel distribution then you can use the method of moments to calculate $F(X)$ values. Moments are statistical descriptors of a data set. The first moment of a data set is the mean; the second moment the standard deviation; the third moment skewness; the fourth kurtosis. To use the formula below you must first find the mean $(\bar{Q})$ and standard deviation σ_Q of your annual maximum data series. The symbol e in the following equation is the base number for natural logarithms or the exponential number (≈ 2.7183).

$$F(X) = e^{-e^{-b(X-a)}}$$

$$a = \bar{Q} - \frac{0.5772}{b}$$

$$b = \frac{\pi}{\sigma_Q \sqrt{6}}$$

With knowledge of $F(X)$ you can find $P(X)$ and the average recurrence interval $(T(X))$ for a certain size of flow: X. The formulae above can be rearranged to give you the size of flow that might be expected for a given average recurrence interval:

$$X = a - \frac{1}{b} \ln \ln \left(\frac{T(X)}{T(X) - 1} \right)$$

In the formula above ln represents the natural logarithm. To find the flow for a 50-year average recurrence interval you must find the natural logarithm of (50/49) and then the natural logarithm of this result.

Using this method it is possible to find the resultant flow for a given average recurrence interval that is beyond the length of your time series. The further away from the length of your time series you move the more error is likely to be involved in the estimate. As a general rule of thumb it is considered reasonable to extrapolate up to twice the length of your streamflow record, but you should not go beyond this.

Low flow frequency analysis

Where frequency analysis is concerned with low flows rather than floods, the data required are an annual minimum series. The same problem is found as for annual maximum series: which annual year to use when you have to assume that the annual minimum flows are independent of each other. At mid-latitudes in the northern hemisphere the calendar year is the most sensible, as you would expect the lowest flows to be in the summer months (i.e. the middle of the year of record). Elsewhere an

Worked example of flood frequency analysis

The data used to illustrate the flood frequency analysis are from the same place as the flow duration curve (the upper Wye catchment in mid-Wales, UK). In this case it is an annual maximum series for the period 1971 until 1997 (inclusive).

In order to establish the best time of year to set a cut-off for the hydrological year all daily mean flows above a threshold value (4.5 m³/s) were plotted against their day number (Figure 7.11). It is clear from Figure 7.11 that high flows can occur at almost any time of the year although at the start and end of the summer (150 = 30 May; 250 = 9 September) there are slight gaps. The hydrological year from June to June is sensible to choose for this example.

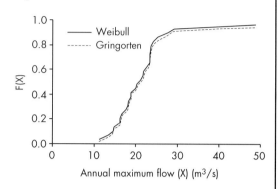

Figure 7.13 Frequency of flows less than X plotted against the X values. The $F(X)$ values are calculated using both the Weibull and Gringorten formulae.

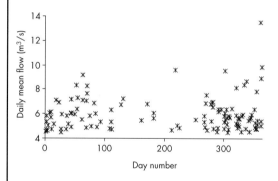

Figure 7.12 Daily mean flows above a threshold value plotted against day number (1–365) for the Wye catchment.

The Weibull and Gringorten position plotting formulae are both applied to the data (see Table 7.3) and the $F(X)$, $P(X)$ and $T(X)$ (average recurrence interval) values calculated. The data look different from those in Figure 7.3, and from the flow duration curve, because they are the peak flow values recorded in each year. This is the peak value of each storm hydrograph, which is not the same as the peak mean daily flow values.

When the Weibull and Gringorten values are plotted together (Figure 7.13) it can be seen that there is very little difference between them.

When the data are plotted with a transformation to fit the Gumbel distribution they almost fit a straight line, suggesting that they do fit a distribution for extreme values such as the Gumbel but that a larger data set would be required to make an absolute straight line. A longer period of records is likely to make the extreme outlier lie further along the x-axis. The plot presented here has transformed the data to fit the Gumbel distribution. Another method of presenting this data is to plot them on Gumbel distribution paper. This provides a non-linear scale for the x-axis based on the Gumbel distribution.

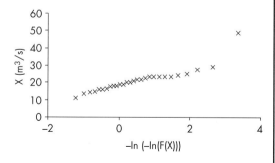

Figure 7.14 Frequency of flows less than a value X. NB The $F(X)$ values on the x-axis have undergone a transformation to fit the Gumbel distribution (see text for explanation).

Table 7.3 Annual maximum series for the Wye (1971–97) sorted using the Weibull and Gringorten position plotting formulae

Rank	X	F(X) Weibull	F(X) Gringorten	P(X)	T(X)
1	11.17	0.03	0.02	0.97	1.04
2	13.45	0.07	0.05	0.93	1.07
3	14.53	0.10	0.09	0.90	1.12
4	14.72	0.14	0.12	0.86	1.16
5	16.19	0.17	0.16	0.83	1.21
6	16.19	0.21	0.19	0.79	1.26
7	16.58	0.24	0.23	0.76	1.32
8	17.57	0.28	0.26	0.72	1.38
9	18.09	0.31	0.29	0.69	1.45
10	18.25	0.34	0.33	0.66	1.53
11	18.75	0.38	0.36	0.62	1.61
12	18.79	0.41	0.40	0.59	1.71
13	20.01	0.45	0.43	0.55	1.81
14	20.22	0.48	0.47	0.52	1.93
15	21.10	0.52	0.50	0.48	2.07
16	21.75	0.55	0.53	0.45	2.23
17	21.84	0.59	0.57	0.41	2.42
18	22.64	0.62	0.60	0.38	2.64
19	23.28	0.66	0.64	0.34	2.90
20	23.36	0.69	0.67	0.31	3.22
21	23.37	0.72	0.71	0.28	3.63
22	23.46	0.76	0.74	0.24	4.14
23	23.60	0.79	0.77	0.21	4.83
24	24.23	0.83	0.81	0.17	5.80
25	25.19	0.86	0.84	0.14	7.25
26	27.68	0.90	0.88	0.10	9.67
27	29.15	0.93	0.91	0.07	14.50
28	48.87	0.97	0.95	0.03	29.00

Table 7.4 Values required for the Gumbel formula, derived from the Wye data set in Table 7.3

Mean ($\bar{Q}$)	Standard deviation (σ_Q)	a value	b value
21.21	6.91	18.11	0.19

Applying the method of moments and Gumbel formula to the data gives some interesting results. The values used in the formula are shown in Table 7.4 and can be easily computed. When the formula is applied to find the flow values for an average recurrence interval of 50 years it is calculated as 39.1 m³/s. This is less than the largest flow during the record which under the Weibull formula has an average recurrence interval of 27 years. This discrepancy is due to the method of moments formula treating the highest flow as an extreme outlier. If we invert the formula we can calculate that a flood with a flow of 48.87 m³/s (the largest on record) has an average recurrence interval of around three hundred years.

analysis of when low flows occur needs to be carried out so that the hydrological year avoids splitting in the middle of a low flow period. In this case $P(X)$ refers to the probability of an annual minimum greater than or equal to the value X. The formulae used are the same as for flood frequency analysis (Weibull etc.).

There is one major difference between flood frequency and low flow frequency analysis which has huge implications for the statistical methods used: there is a finite limit on how low a flow can be. In theory a flood can be of infinite size, whereas it is not possible for a low flow to be less than zero (negative flows should not exist in fresh water hydrology). This places a limit on the shape of a probability distribution, effectively truncating it on the left-hand side (see Figure 7.15).

The statistical techniques described on pp. 98–103 (for flood frequency analysis) assume a full log-normal distribution and cannot be easily applied for low flows. Another way of looking at this problem is shown in Figure 7.16 where the probabilities calculated from the Weibull formula are plotted against the annual minimum flow values. The data fit a straight line, but if we extrapolate the line further it would intersect the x-axis at a value of approximately 0.95. The implication from this is that there is a 5 per cent chance of having a flow less than zero (i.e. a negative flow). The way around this is to fit an exponential rather than a straight line to the data. This is easy to do by eye but complicated mathematically. It is beyond the level of this text to describe the technique here (see Shaw [1994] or Wang and Singh [1995] for more detail). Gordon *et al.* (1992) provide a simple method of overcoming this problem, without using complicated line-fitting procedures.

Limitations of frequency analysis

As with any estimation technique there are several limitations in the application of frequency analysis; three of these are major:

1   The estimation technique is only as good as the

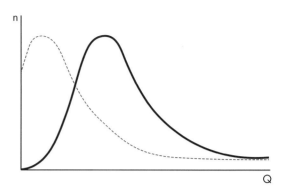

Figure 7.15 Two probability density functions. The usual log-normal distribution (solid line) is contrasted with the truncated log-normal distribution (dashed line) that is possible with low flows (where the minimum flow can equal zero).

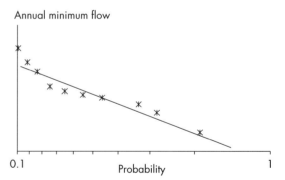

Figure 7.16 Probability values (calculated from the Weibull sorting formula) plotted on a log scale against values of annual minimum flow (hypothetical values).

streamflow records that it is derived from. Where the records are short or of dubious quality very little of worth can be achieved through frequency analysis. As a general rule of thumb you should not extrapolate average recurrence intervals beyond twice the length of your data set. There is a particular problem with flood frequency analysis in that the very large floods can create problems for flow gauges and therefore this extreme data may be of dubious quality.

2   The assumption is made that each storm or low flow event is independent of another used in the data set. This is relatively easy to guard against in annual maximum (or minimum) series, but more difficult for a peak threshold series.

3   There is an inherent assumption made that the hydrological regime has remained static during the complete period of record. This may not be true where land use, or climate change, has occurred in the catchment (see Chapter 9).

SUMMARY

The analysis of streamflow records is extremely important in order to characterise the flow regime for a particular river. Hydrograph analysis involves dissecting a hydrograph to distinguish between stormflow and baseflow. This is often a precursor to using the unit hydrograph, a technique using past stormflow records to make predictions on the likely form of future storm events. Flow duration curves are used to look at the overall hydrology of a river – the percentage of time a river has an average flow above or below a certain threshold. Frequency analysis is used to look at the average return period of a rare event (or the probability of a certain rare event occurring), whether that be extremes of flooding or low flow. Each of the methods described in this chapter has a distinct use in hydrology and it is important that practising hydrologists are aware of their role.

ESSAY QUESTIONS

1   **Describe the technique of hydrograph separation and explain why it is carried out.**

2   **Outline the limitations of the unit hydrograph when used as a predictive tool and attempt to explain its success despite these limitations.**

3   **Describe the types of information that can be derived from a flow duration curve and explain the use of that information in hydrology.**

4   **Explain why interpretation of flood (or low flow) frequency analysis may be fraught with difficulty.**

5   **Describe the type of data (and measurement equipment) required for describing the general hydrology of a small catchment near you.**

FURTHER READING

Dingman, S.L. (1994) *Physical hydrology*. Macmillan, New York.
A high level text with good detail on analytical techniques.

Shaw, E.M. (1994) *Hydrology in practice* (3rd edition). Chapman and Hall, London.
An engineering text with good detail on analytical techniques.

8

WATER QUALITY

LEARNING OBJECTIVES

When you have finished reading this chapter you should have:

- An understanding of water quality as an issue in hydrology and how it ties into water quantity.
- A knowledge of the main parameters used to assess water quality and what affects their levels in a river.
- A knowledge of the measurement techniques and sampling methodology for assessing water quality.
- A knowledge of techniques used to control water pollution and manage water quality.

This chapter identifies the different types of pollutants that can be found in a river system and describes the major sources of them, especially for where elevated levels may be found. The methods that can be used to measure water quality parameters are outlined. The chapter finishes with a description of the management techniques that can be used to control water quality in a river catchment.

Traditionally hydrology has been interested purely in the amount of water in a particular area: water quantity. If, however, we take a wider remit for hydrology – to include the availability of water for human consumption – then issues of water quality are of equal importance to quantity. There are three strong arguments as to why hydrology should consider water quality an area worthy of study.

1 The interlink between water quality and quantity. Many water quality issues are directly linked to the amount of water available for dilution and dispersion of pollutants, whether they be natural or anthropogenic in source. It is virtually impossible to study one without the other. An example of this is shown in the Case Study of the river Thames through London (p. 107).

2 The interlink between hydrological processes and water quality. The method by which pollutants transfer from the land into the aquatic environ-

ment is intrinsically linked with the hydrological pathway (i.e. the route by which the water moves from precipitation into a stream), and hence the hydrological processes occurring. A good example of this is in Heppell *et al*. (1999) where the mechanisms of herbicide transport from field to stream are linked to hydrological pathways in a clay catchment.

3 Employment of hydrologists. It is rare for someone employed in the water industry to be entirely concerned with water quantity, with no regard for quality issues. The maintenance of water quality is not just for drinking water (traditionally an engineer's role) but at a wider scale can be for maintaining the **amenity value** of rivers and streams.

Case study

THE RIVER THAMES THROUGH LONDON: COMPLETE HYDROLOGY

The river Thames as it flows through London is one of the great tourist sights of Europe. It is an integral part of London, not just for its scenic attraction but also as a transport route right into the heart of a modern thriving city. The river has also a large part to play in London's water resources, both as source of drinking water and a disposal site for waste.

London has a long history of water-quality problems on the Thames, but it has not always been so. Prior to the nineteenth century domestic waste from London was collected in cesspools and then used as fertiliser on agricultural land (hence the use of the term 'sewage farm' for sewage treatment stations). The Thames had a fish population, and salmon from the river were sold for general consumption. With the introduction of compulsory water closets (1843) and the rise in factory waste during the Industrial Revolution things started to change dramatically for the worse during the nineteenth century. The majority of London's waste went through poorly constructed sewers (often leaking into shallow aquifers which supplied much drinking water) straight into the Thames without any form of treatment. In 1854 there was an outbreak of cholera in London that resulted in up to 10,000 deaths. In a famous epidemiological study Dr John Snow was able to link the cholera to sewage pollution in water drawn from shallow aquifers. The culmination of

this was 'the year of the great stink' in 1856. The smell of untreated waste in the Thames was so bad that disinfected sheets had to be hung from windows in the Houses of Parliament to lessen discomfort for the lawmakers of the day. In the best NIMBY ('not in my backyard') tradition this spurred parliament into action and in the following decade, radical changes were made to the way that London used the river Thames. Water abstraction for drinking was only allowed upstream of tidal limits and London's sewage was piped downstream to Beckton where it was discharged (still untreated) into the Thames on an ebb tide.

The result of these reforms was a radical improvement of the river water quality through central London; but there was still a major problem downstream of Beckton. The improvements were not to last, however, as by the middle of the twentieth century the Thames was effectively a dead river (i.e. sustained no fish population and had a dissolved oxygen concentration of zero for long periods of the summer). This was the result of several factors: a rapidly increasing population, increasing industrialisation, a lack of investment in sewage treatment, and bomb damage during the Second World War.

Since the 1950s the Thames has been steadily improving. Now there is a maintained fish population and migratory salmon can move up the

Table 8.1 Comparison of rivers flowing through major cities

River	Mean annual flow (m³/s)	City on river or estuary	Population in metropolitan area (million)
Thames	82	London	11.2
Seine	268	Paris	11.3
Hudson	387	New York	29.5
Sacramento/San Joaquin	860	San Francisco	7.0
Rhine	2,219	Rotterdam	1.1
Paraná/Uruguay	22,000	Buenos Aires	11.6

Source: Flow data from Global Runoff Data Centre

Thames. This improvement has been achieved through an upgrading of the many sewage treatment works that discharge into the Thames and its tributaries. The Environment Agency has much to do with the management of the lower Thames and proudly proclaims that the Thames 'is one of the cleanest metropolitan rivers in the world'. How realistic is this claim?

There is no doubt that the Thames has been transformed remarkably from the 'dead' river of 60 years ago into something far cleaner, but there are two problems remaining for the management of the Thames through London, and for one of these nothing can be done.

- The Thames is a relatively small river that does not have the flushing potential of other large rivers; therefore it cannot cleanse itself very easily.
- The sewer network underneath London has not been designed for a large modern city and cannot cope with the strains put on it.

At Westminster (in front of the Houses of Parliament) the Thames is over 300 m wide; this is confined from the width of 800 m evident during Roman times. This great width belies a relatively small flow of fresh water. It appears much larger than general flow statistics would suggest because of its use for navigation and the tidal influence. The average flow rate for the Thames is 53 cumecs, rising to 130 cumecs under high flows.

In Table 8.1 this is compared with rivers that flow through other major cities. The effect of the small flow in the Thames is that it does not have great flushing power. It may take a body of water up to 3 months (during the summer) to move from west London to the open sea. On each tide it may move up to 14 km in total but this results in less than a kilometre movement downstream. If this body of water is polluted in some way then it is not receiving much dilution or dispersion during the long trip through London.

The second important factor is the poor state of London's sewers. Prior to Sir Joseph Bazalgette's sewer network of 1864 the old tributaries of the Thames acted as sewers, taking the waste directly to the Thames. Bazalgette's grand sewerage scheme intercepted these rivers and transported the sewage through a large pipe to east London. This system still exists today. The actual sewerage network is very well built and still works effectively. The problem is that it is not able to cope with the volume of waste expected to travel through it. This is particularly so when it rains, as this is a combined stormwater–sewage network. The original tributaries of the Thames, such as the Fleet, still exist under London and any storm runoff is channelled into them. When the volume of stormwater and sewage is too great they act as overspills and take the untreated sewage directly into the Thames. This is a particular problem during summer storms when the volume of water flowing down the Thames is low and cannot dilute the waste effectively. To combat this problem Thames Water Utilities (part of the private company that treats London's sewage) operate two boats specially designed to inject oxygen directly into the water. These boats can float with a body of sewage-polluted water, injecting oxygen so that the dissolved oxygen level does not reach levels that would be harmful to fish and other aquatic creatures. To help in the tracking of a polluted body of water there are water-quality monitoring stations attached to bridges over the Thames. These stations measure temperature, dissolved oxygen concentration and electrical conductivity at 15-minute intervals and are monitored by the Environment Agency as they are received in real time at the London office.

In addition to the oxygen-injecting boats, there is tight water-quality management for the river Thames through London. This is operated by the Thames Estuary Partnership, a group of interested bodies including the Environment Agency. Their remit includes other factors such as protecting London from flooding (using the Thames Barrier), but has factors in it such as higher effluent standards for sewage treatment works during the summer. The emphasis is on flexibility in their management of the Thames. There is no question that the River Thames has improved from 50 years ago. In many respects it is a river transformed, but it still has major water-quality problems such as you would expect to find where a small river is the receptacle for the treated waste of over 10 million people. The water-quality management of a river like the Thames needs consideration of many facets of hydrology: understanding pollutants, knowledge of stormflow peaks from large rainfall events, and streamflow statistics.

It is easy to think of water quality purely in terms of pollution: waste substances entering a river system as a result of human activity. This is an important issue in water-quality analysis but is by no means the only one. One of the largest water quality issues is the amount of suspended sediment in a river, which is frequently a completely natural process. Suspended sediment has severe implications for the drinking-water quality of a river, but also for other hydrological concerns such as reservoir design and aquatic flora and fauna. As soon as a river is dammed the water velocity will slow down. Simple knowledge of the **Hjulstrom curve** (see Figure 8.1) tells us that this will result in the deposition of suspended sediment. That deposition will eventually reduce the capacity of the reservoir held behind the dam. In high-energy river systems, for even a very large reservoir, a dramatic reduction in capacity can take place within two to four decades (e.g. the Roxburgh dam on the Clutha river, South Island of New Zealand). It is critically important for a hydrologist involved in reservoir design to have some feeling for the quantities of suspended sediment so that the lifespan of a reservoir can be calculated.

Spatial variations in water quality may be influenced by many different environmental factors (e.g. climate, geology, weathering processes, vegetation cover and anthropogenic). Often it is a combination of these factors that makes a particular water-quality issue salient for a particular area. An example of this is acid rain as a particular problem for north-eastern North America and Scandinavia. The sources of

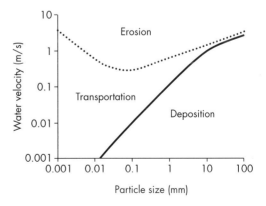

Figure 8.1 The Hjulstrom curve relating stream velocity to the erosion/deposition characteristics for different sized particles (x-axis).

the acid rain are fossil-fuel-burning power stations and industry. It becomes a particular problem in these areas for a number of reasons: it is close to the sources of acid rain; high rainfall contributes a lot of acid to the soil; the soils are derived from a very old geology and are often thin after extensive glaciation; the soils are heavily leached (have had a lot of water passing through them over a long time period) and have a low buffering capacity (see pp. 113–14) for the acid rain. This combination of influences means that the water in the rivers has a low **pH**, and – of particular concern to gill-bearing aquatic fauna – has a high dissolved aluminium content (from the soils).

Having argued for the role of natural water-quality issues to be considered seriously the reader will find that the majority of this chapter deals with human-induced water-quality issues. This is an inevitable response to the world we live in where we place huge pressures on the river systems as repositories of waste products. It is also important to study these issues because they are something that humans can have some control over, unlike many natural water-quality issues.

Before looking at the water-quality issues of substances within a river system it is worth considering how they reach a river system. In studying water pollution it is traditional to differentiate between *point source* and *diffuse* pollutants. As the terminology suggests point sources are discrete places in space (e.g. a sewage treatment works) where pollutants originate. Diffuse sources are spread over a much greater land area and the exact locations cannot be specified. Examples of this type of pollutant are excess fertilisers and pesticides from agricultural production. The splitting of pollutants into diffuse and point sources has some merit for designing preventative strategies but like most categorisations there are considerable overlaps. Although a sewage treatment works can be thought of as a point source when it discharges effluent into a stream it has actually gathered its sewage from a large diffuse area. If there is a particular problem with a sewage treatment works effluent, it may be a result of accumulated diffuse source pollution rather than the actual sewage treatment works itself.

A more useful categorisation of water pollutants is to look at their impacts on the river system. In this way we can differentiate between three major types of pollutants.

- *Toxic compounds*, which cause damage to biological activity in the aquatic environment.
- *Oxygen balance affecting compounds*, which either consume oxygen or inhibit the transfer of oxygen between air and water. This would also include thermal pollution as warm water does not hold as much dissolved oxygen as cold water (see p. 114).
- *Suspended solids* – the inert solid particles suspended in the water.

Whether we approve or not, rivers are receptacles for large amounts of waste produced by humans. Frequently this is deliberate and is due to the ability of rivers to cope with waste through the three ds: *d*egradation; *d*ilution; and *d*ispersion. Just how quickly these processes operate is dependent on different factors. These factors include the pollutant load already present in the river, the temperature and pH of the water, the amount of water flowing down the river, and the mixing potential of the river. The last two of these are river flow characteristics that will in turn be influenced by the time of year, the nature of flow in the river (e.g. the shape of the flow duration curve), and the velocity and turbulence of flow. This demonstrates the strong interrelationship that exists between water quality and water quantity in a river system.

One remarkable feature about rivers is that given enough time and a reasonable pollution loading, rivers will recover from the input of many pollutant types. That is not to say that considerable harm cannot be done through water pollution incidents, but by and large the river system will recover so long as the pollution loading is temporary. An example of this can be seen in the **oxygen sag curve** (see Figure 8.2) that is commonly seen below point sources of organic pollution (e.g. sewage effluent). The curve shows that upon entering the river there is an instant drop in dissolved oxygen content. This is caused by bacteria and other micro-organisms

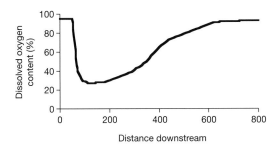

Figure 8.2 Hypothetical dissolved oxygen sag curve. The point at which the curve first sags is the point source of an organic pollutant. The distance downstream has no units attached as it will depend on the size of the river.

in the river feeding on the organic matter in the stream and using any available dissolved oxygen. This would have a severe impact on any aquatic fauna unable to move away from this zone of low dissolved oxygen. As the pollutant load moves downstream the degradation, dilution and dispersal starts to take effect and oxygen levels start to recover in the river. The shape of the curve, especially the distance downstream until recovery, is highly dependent on the flow regime of the receiving river. A fast flowing, readily oxygenated stream will recover much faster than a slow-moving river. Large rivers will have a faster recovery time (and the depth of sag will be less) than small streams, due to the amount of dilution occurring.

WATER-QUALITY PARAMETERS

To analyse the water quality within a river consideration has to be given to what type of test may be carried out and the sampling pattern to be used. There are numerous parameters that can be measured, and each is important for the part they play in an overall water-quality story. It is not necessary to measure them all for a single water-quality analysis study, instead the relevant parameters for a particular study should be identified. This can be done using a priori knowledge of the water-quality issues being studied. To aid in this different

parameters are discussed here with respect to their source; what type of levels might be expected in natural rivers; and the impact they have on a river. In addition to this various parameters are described which provide a general overview of water quality without necessarily measuring a concentration of a certain substance.

The first distinction that can be made is between physical and chemical parameters. With chemical parameters it is the concentration of a particular chemical substance that is being assessed. With physical parameters it is a physical measurement being made, normally measuring the amount of something within a water sample.

Physical parameters

Temperature

The temperature of water in a river is an important consideration for several reasons. The most important feature of temperature is the interdependence it has with dissolved oxygen content (see p. 114). Warm water holds less dissolved oxygen than colder water. The dissolved oxygen content is critical in supporting aquatic fauna, so temperature is also indirectly important in this manner. In addition to this, water temperature is also a controlling factor in the rate of chemical reactions occurring. Warm water will increase the rate of many chemical reactions occurring in a river, and it is able to dissolve more substances. This is due to a weakening of the hydrogen bonds and a greater ability of the bipolar molecules to surround anions and cations.

Warm water may enter a river as thermal pollution from power stations and other industrial processes. In many power stations (gas, coal and nuclear) water is used as a coolant in addition to the generation of steam to drive turbines. Because of this, power stations are frequently located near a river or lake to provide the water source. It is normal for the power stations to have procedures in place so that hot water is not discharged directly into a river; however, despite the cooling processes used, the water is frequently 1–2°C degrees warmer on discharge. The impact that this has on a river system will be dependent on the river size.

Dissolved solids

In the first chapter, the remarkable ability of water to act as a solvent was described. As water passes through a soil column or over a soil surface it will dissolve many substances attached to the soil particles. Equally water will dissolve particles from the air as it passes through the atmosphere as rain. The amount of dissolved substances in a water sample is referred to as the **total dissolved solids (TDS)**. The higher the level of TDS the more contaminated a water body may be, whether that be from natural or anthropogenic sources. Meybeck (1981) estimates that the global average TDS load in rivers is around 100 mg/l, but it may rise considerably higher (e.g. the Colorado river has a TDS of 703 mg/l).

Electrical conductivity

A similar measurement to TDS is provided by the electrical conductivity. The ability of a water sample to transmit electrical current (its conductivity) is directly proportional to the concentration of dissolved ions. Pure, distilled water will still conduct electricity but the more dissolved ions in a water sample the higher its electrical conductivity. This is a straight-line relationship, so the following equation can be derived:

$$K = \frac{\text{Conductivity}}{\text{TDS}} \text{ or TDS} = \frac{\text{Conductivity}}{K}$$

This relationship gives a very good surrogate measure for TDS. The K term is a constant (usually between 0.55 and 0.75) that can be estimated by taking several measurements of conductivity with differing TDS levels. Conductivity is a simple measurement to take as there are many robust field instruments that will give an instant reading. This can then be related to the TDS level at a later stage. Electrical conductivity is measured in Siemens per metre, although the usual expression is microsiemens per centimetre (μS/cm). Rivers normally have a conductivity between 10 and 1,000 μS/cm.

Suspended solids

The amount of suspended solid has been highlighted at the start of this chapter as a key measure of water quality. The carrying of suspended sediment in a river is part of the natural erosion process. The sediment will be deposited at any stage when the river velocity drops and conversely it will be picked up again with higher river velocities (see Figure 8.1). In this manner the **total suspended solids (TSS)** load will vary in space and time. The amount of TSS in a river will affect the aquatic fauna, because it is difficult for fish and invertebrates to breed in an environment of high sediment. Suspended sediment is frequently inert, as in the case of most clay and silt particles, but it can be organic in content and therefore have an oxygen demand.

TSS is expressed in mg/l for a water sample but frequently uses other units when describing sediment load. Table 8.2 shows some values of sediment discharge (annual totals) and calculates an average TSS from the data. It is remarkable to see the data in this form, enabling contrast to be drawn between the different rivers. Although the Amazon delivers a huge amount of sediment to the oceans it has a relatively low average TSS, a reflection of the extremely high discharge. In contrast to this the Huanghe river (sometimes referred to as the Yellow river due to the high sediment load) is virtually a soup! It must be noted that these are average values over a year and that the TSS will vary considerably during an annual cycle (the TSS will rise considerably during a flood).

Turbidity

A similar measure to TSS is the **turbidity**: a measure of the cloudiness of water. The cloudiness is caused by suspended solids and gas bubbles within the water sample, so the two are directly related. Turbidity is measured as the amount of light

Table 8.2 Sediment delivery, total river discharge (averaged over several years) and average total suspended solids (TSS) for selected large river systems

River (country)	Sediment discharge (10^3 tonnes/yr)	Discharge (km³/yr)	Average TSS (mg/l)
Zaïre (Zaïre)	43,000	1,250	0.03
Amazon (Brazil)	900,000	6,300	0.14
Danube (Romania)	67,000	206	0.33
Mississippi (USA)	210,000	580	0.36
Murray (Australia)	30,000	22	1.36
Ganges-Brahmaputra (Bangladesh)	1,670,000	971	1.72
Huanghe or Yellow (China)	1,080,000	49	22.04

Source: Data from Milliman and Meade (1983)

scattered by the suspended particles in the water. A beam of light of known luminosity is shone through a sample and the amount reaching the other side is measured. This is compared to a standard solution of formazin. The units for turbidity are either FTU (formazin turbidity units) or NTU (normalised turbidity units); they are identical. Turbidity is a critical measure of water quality for the same reasons as TSS. It is a simpler measurement to make, especially in the field, and therefore it is sometimes used as a surrogate for TSS.

Chemical parameters

pH

Chemists think of water as naturally disassociating into two separate ions: the hydroxide (OH⁻) and hydrogen (H⁺) ions.

$$H_2O \rightleftharpoons OH^- + H^+$$

The acidity of a water is given by the hydrogen ion, and hence pH (the measure of acidity) is a measure of the concentration of hydrogen ions present. In fact it is the log of the inverse concentration of hydrogen ions.

$$pH = \log \frac{1}{\left[H^+ \right]}$$

This works out on a scale between 1 and 14, with 7 being neutral. A pH value less than 7 indicates an acid solution; greater than 7 a basic solution (also called alkaline). It is important to bear in mind that because the pH scale is logarithmic (base 10) a solution with pH value 5 is ten times as acidic as one with pH value 6.

In natural waters the pH level may vary considerably. Rainwater will naturally have a pH value less than 7, due to the absorption of gases such as carbon dioxide by the water. This forms a weak carbonic acid, increasing the concentration of hydrogen ions in solution. The normal pH of rainfall is somewhere between 5 and 6 but may drop as low as 4, particularly if there is industrial air pollution nearby. Zhao and Sun (1986) report a pH value of 4.02 in Guiyang city, China, during 1982.

Acidic substances may also be absorbed easily as water passes through a soil column. A particular example of this is water derived from peat, which will absorb organic substances. These form organic acids, giving peat-derived water a brown tinge and a low pH value. At the other end of the spectrum rivers that drain carbonate-rich rocks (e.g. limestone and chalk), have a higher pH due to the dissolved bicarbonate ions and base cations (e.g. Ca^{2+}).

The pH value of rivers is important for the aquatic fauna living within them. The acidity of a river is an important control for the amount of dissolved ions present, particularly metal species.

The more acidic a river is the more metallic ions will be held in solution. For fish it is often the level of dissolved aluminium that is critical for their survival in low pH waters. The aluminium is derived from the breakdown of alumino-silicate minerals in clay, a process that is enhanced by acidic water. Water with a pH between 6 and 9 is unlikely to be harmful to fish. Once it drops below 6 it becomes harmful for breeding, and salmonid species (e.g. trout and salmon) cannot survive at a pH lower than 4. Equally a pH higher than 10 is toxic to most fish species (Alabaster and Lloyd, 1980).

Mention needs to be made of the confusing terminology regarding **alkalinity**. Alkalinity is a measure of the capacity to absorb hydrogen ions without a change in pH (Viessman and Hammer, 1998). This is influenced by the concentration of hydroxide, bicarbonate or carbonate ions. In water-quality analysis the term 'alkalinity' is used almost exclusively to refer to the concentration of bicarbonate (HCO_3^-) ions because this is the most variable of the three. The bicarbonate ions are derived from the percolation of water through calcareous rocks (e.g. limestones or chalk). It is important to know their concentration for the buffering of pH and for issues of water hardness. The buffering capacity of soils, and water derived from soils, is an important concept in water quality. The buffering capacity of a solution is the ability to absorb acid without changing the pH. This is achieved through a high base cation load or high bicarbonate load. This is why soils derived from limestone and chalk have less problem with acid rain.

Dissolved oxygen

Dissolved oxygen is vital to any aquatic fauna that use gills to breath. Salmonid species of fish require a dissolved oxygen content greater than 5 mg/l, whereas coarse fish (e.g. perch, pike) can survive in levels as low as 2 mg/l. The dissolved oxygen content is an important factor in the way we taste water. Water saturated in oxygen tastes fresh to human palates, hence drinking water is almost

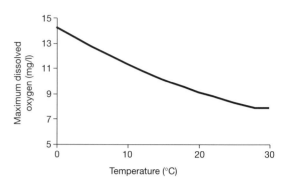

Figure 8.3 Relationship between maximum dissolved oxygen content (i.e. saturation) and temperature.

always oxygenated before being sent through a pipe network to consumers.

There are two methods by which dissolved oxygen content is considered: percentage saturation and concentration (mg/l). These two measures are interrelated through temperature, as the dissolved oxygen content of water is highly temperature dependent (see Figure 8.3).

Biochemical oxygen demand

One of the key water-quality parameters is the five-day biochemical oxygen demand test (sometimes referred to as the **biological oxygen demand** test, or BOD_5). This is a measure of the oxygen required by bacteria and other micro-organisms to break down organic matter in a water sample. It is an indirect measure of the amount of organic matter in a water sample, and gives an indication of how much dissolved oxygen could be removed from a stream due to the organic matter.

The test is simple to perform and easily replicable. A sample of water needs to be taken, placed in a clean, darkened glass bottle and left to reach 20°C. Once this has occurred the dissolved oxygen content should be measured (as a concentration). The sample should then be left at 20°C for five days in a darkened environment. After this the dissolved oxygen content should be measured again. The difference between the two dissolved oxygen readings is the BOD_5 value. Over an extended period the dissolved oxygen content of a polluted water sample will look something like that shown in Figure 8.4. In this case the dissolved oxygen content has dropped from 9.0 on day one to 3.6 on day five, giving a BOD_5 value of 5.4 mg/l. After a long period of time (normally more than five days) oxygen will start to be consumed by nitrifying bacteria. In this case the bacteria will be consuming oxygen to turn nitrogenous compounds (e.g. ammonium ions) into nitrate. In order to be sure that nitrifying bacteria are not adding to the oxygen demand a suppressant (commonly allyl thourea or ATU) is added. This ensures that all the oxygen demand is from the decomposition of organic matter. The use of a five-day period is another safeguard, as due to the slow growth of nitrifying bacteria their effect is not noticeable until 8–10 days (Tebbutt, 1993). There is an argument to be made saying that it does not matter which bacteria are causing the oxygen demand, the test should be looking at all oxygen demand over a five-day period and therefore there is no need to add ATU.

In some cases, particularly when dealing with waste water, the oxygen demand will be higher than total saturation. In this case the sample needs to be diluted with distilled water. The maximum dissolved oxygen content at 20°C is 9.1 mg/l, so any water sample with a BOD_5 value higher than 9 will require dilution. After the diluted test a calculation needs to be performed to find the actual oxygen demand. If you have diluted the sample by half then you need to double your measured BOD_5 value, and so on.

A normal unpolluted stream should have a BOD_5 value of less than 5 mg/l. Untreated sewage is somewhere between 220 and 500 mg/l; while milk has a BOD_5 value of 140,000 mg/l. From these values it is possible to see why a spillage of milk into a stream can have such detrimental effects on the aquatic fauna. The milk is not toxic in its own right, but bacteria consuming the milk will strip the water of any dissolved oxygen and therefore deprive fish of any.

There are three reasons why BOD_5 is such a crucial test for water quality:

- It has huge implications for aquatic fauna that depend on the dissolved oxygen to survive.
- It is an indirect measure of the amount of organic matter in the water sample.
- It is the most frequently measured water quality test and has become a standard measure; this means that there are plenty of data to compare readings against.

It also important to realise that BOD is not a direct measure of pollution; rather, it measures the effects of pollution. It also should be borne in mind that there may be other substances present in your water sample that inhibit the natural bacteria (e.g. toxins). In this case the BOD_5 reading may be low despite a high organic load.

Trace organics

Over six hundred organic compounds have been detected in river water, mostly from human activity (Tebbutt, 1993). Examples include benzene, chlorophenols, pesticides, trihalomethanes, and polynuclear aromatic hydrocarbons (PAH). These would normally be found in extremely low concentrations but do present significant health risks over the long term.

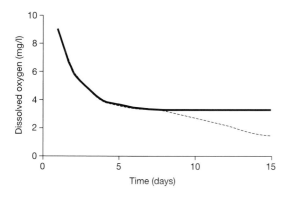

Figure 8.4 Dissolved oxygen curve. The solid line indicates the dissolved oxygen content decreasing due to organic matter. The dashed line shows the effect of nitrifying bacteria.

Table 8.3 Percentage of water resources with pesticide concentrations regularly greater than 0.1 μg/l (European Union drinking water standard) for selected European countries

Country	Surface water (%)	Groundwater (%)
Belgium	100	5.2
Denmark	n/a	8.9
Germany	0.0	0.0
Netherlands	50.0	5.0
UK	77.0	6.0

Source: Data from Eureau (2001)

The data for pesticide concentrations (see Table 8.3) in European water resources show that it is a significant problem. This indicates that all water extracted from surface water supplies in Belgium (supplies approximately 30 per cent of the Belgian population) will require pesticide removal before reticulation to customers (Eureau, 2001). Although Germany appears to have no pesticide problem, 10 per cent of its surface water resources occasionally have pesticide levels greater than 0.1 μg/l and 90 per cent have pesticides in concentrations less than 0.1 μg/l (but still present) (Eureau, 2001).

Some of the trace organic compounds accumulate through the food chain so that humans and other species that eat large aquatic fauna may be at risk. Of particular concern are endocrine disrupting chemicals (EDCs), which have been detected in many rivers. These chemicals, mostly a by-product of industrial processes, attack the endocrine system of humans and other mammals, affecting hormone levels. Some chemicals (e.g. DDT) have the ability to mimic the natural hormone oestrogen. Because oestrogen is part of the reproductive process these chemicals have the potential to affect reproductive organs and even DNA. Studies have shown high levels of oestrogen-mimicking compounds in sewage effluent (Montagnani *et al*., 1996) and that male fish held in cages at sewage effluent discharge sites can develop female sexual organs (Jobling and Sumpter, 1993).

Trace organics can be detected using gas chromatography, although this is made difficult by the sheer number of compounds to be detected. They are removed from drinking water supplies using activated carbon filters, or sometimes oxidation by ozone.

Nitrogen compounds

Nitrogen exists in the freshwater environment in four main forms:

- organic nitrogen – proteins, amino acids and urea
- ammonia – either as free ammonia (NH_3) or the ammonium ion (NH_4^+)
- nitrite (NO_2^-)
- nitrate (NO_3^{2-}).

If organic nitrogen compounds enter a river (e.g. in untreated sewage) then an oxidation process called nitrification takes places. An approximation of the process is outlined below:

$$\text{Organic N} + O_2 \rightarrow NH_3/NH_4^+ + O_2 \rightarrow NO_2^- + O_2 \rightarrow NO_3^{2-}$$

For this to occur there must be nitrifying bacteria and oxygen present. This is one of the main processes operating in a sewage treatment works (see pp. 123–125) – the breakdown of organic nitrogenous compounds into a stable and relatively harmless nitrate. There are two problems with this process occurring in the natural river environment. First, there is the oxygen demand created by the nitrification process. Second, the intermediate ammonia stage is highly toxic, even in very low concentrations. Under extremely low dissolved oxygen concentrations (less than 1 mg/l) the nitrification process can be reversed, at least in the first stage. In this case nitrates will turn into nitrite and oxygen will be released. Unfortunately, this is not a ready means for re-oxygenating a river as by the time the dissolved oxygen level has dropped to 1 mg/l the fish population will have died or moved elsewhere.

The levels of nitrate in a water sample can be expressed in two different ways: absolute nitrate concentration, or the amount of nitrogen held as nitrate (normally denoted as NO_3—N). The two are related by a constant value of approximately 4.4. As an example the World Health Organisation recommended that the drinking water standard for nitrate in drinking water be 45 mg/l. This can also be expressed as 10 mg/l NO_3—N.

As indicated above, one source of nitrate is from treated sewage. A second source is from agricultural fertilisers. Farmers apply nitrate fertilisers to enhance plant growth, particularly during the spring. Plants require nitrogen to produce green leaves, and nitrates are the easiest form to apply as a fertiliser. This is because nitrates are extremely soluble and can easily be taken up by the plant through its root system. Unfortunately this high solubility makes them liable to be flushed through the soil water system and into rivers. To make matters worse a popular fertiliser is ammonium nitrate – $(NH_4)_2NO_3$. This has the added advantage for the farmer of three nitrogen atoms per molecule. It has the disadvantage for the freshwater environment of extremely high solubility and providing ammonium ions in addition to nitrate. The application of nitrate fertilisers is most common in areas of intensive agricultural production such as arable farming.

Another source of nitrates in river systems is from animal wastes, particularly in dairy farming where slurry is applied to fields. This is organic nitrogen (frequently a high urea content from urine) which will break down to form nitrates. This is part of the nitrification process described earlier.

A fourth source of nitrates in river systems is from plants that capture nitrogen gas from the air. This is not strictly true, as it is actually bacteria such as Rhizobium, attached to a plants roots, that capture the gaseous nitrogen and turn it into water-soluble forms for the plants to use. Not all plants have this ability; in agriculture it is the legumes, such as clovers, lucerne (or alfalfa), peas, and soy beans, that can gain nitrogen in this way. Once the nitrogen is in a soluble form it can leach through to a river system in the same way that fertilisers do. Over a summer period the nitrogen levels in a soil build up and then are washed out when autumn and winter rains arrive. This effect is exacerbated by ploughing in the autumn, which releases large amounts of soil-bound nitrogen.

There is one other source of nitrates in rivers: atmospheric pollution. Nitrogen gas (the largest constituent of the atmosphere) will combine with oxygen whenever there is enough energy for it to do so. This energy is readily supplied by combustion engines (cars, trucks, industry, etc.) producing various forms of nitrogen oxide gases (often referred to as NO_x gases). These gases are soluble to water in the atmosphere and form nitrites and nitrates in rainwater. This is not a well-studied area and it is difficult to quantify how much nitrogen reaches rivers from this source.

The different sources of nitrate in a river add together to give a cycle of levels to be expected in a year. Figure 8.5 shows this cycle over a three-year period on the river Lea, south-east England. The low points of nitrate levels correspond to the end of a summer period, with distinct peaks being visible over the autumn to spring period, particularly in the spring. The Lea is a river that has intensive arable agriculture in its upper reaches, but also a significant input from sewage effluent. At times during the summer months the Lea can consist of completely recycled water, and the water may have been through more than one sewage works. This gives a background nitrate level, but it is perhaps surprising that the summer levels of nitrate are not higher, compared to the winter period. Partly this can be attributed to the growth of aquatic plants in

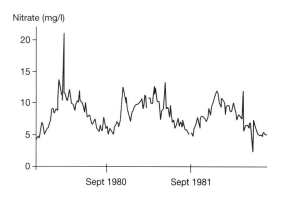

Nitrate (mg/l)

Figure 8.5 Nitrate levels in the river Lea, England. Three years of records are shown: from September 1979 until September 1982.
Source: Data from the Environment Agency

the summer, which remove nitrate from the water. The peaks over the autumn–spring period are as a result of agricultural practices discussed above. The example given here is specific to the south-east of England; in different parts of the world the cycles will differ in timing and extent.

Nitrates are relatively inert and do not create a major health concern. An exception to this is methaemoglobinaemina ('blue baby syndrome'). Newborn babies do not have the bacteria in their stomach to deal with nitrates in the same manner as older children and adults. In the reducing surroundings of the stomach the nitrate is transformed into nitrite that then attaches itself to the haemoglobin molecule in red blood cells, preferentially replacing oxygen. This leads to a reduction in oxygen supply around the body, hence the name 'blue baby syndrome'. In reality methaemoglobinaemina is extremely rare, possibly coming from nitrate-polluted well supplies but not mains-supplied drinking water. The drinking water limit for the European Union is 50 mg/l of nitrate (44 mg/l in the USA). In rivers it is rare to have nitrate values as high as this. In a study of streams draining intensively dairy-farmed land in the North Island of New Zealand, Rodda *et al.* (1999) report maximum nitrate levels of 26.4 mg/l. These are reported as being 'very high by New Zealand standards'

(Rodda *et al.*, 1999: 77). In Figure 8.5 the peak nitrate level for the river Lea in England is 21 mg/l, with the norm being somewhere between 5 and 10 mg/l.

The biggest concern with nitrates in a river system is **eutrophication**. In exactly the same way that the nitrogen enhances the growth of land-based plants, it will also boost the growth of aquatic plants, including algae. This creates a problem of over-production of plant matter in river systems. This is discussed in more detail on pp. 122–123.

Phosphates

Phosphorus can be found in three different forms: orthophosphate, polyphosphate (both normally dissolved) and organic phosphate (bound to organic particles). The ratio of different forms of phosphorus in a water sample is highly pH dependent (Chapman, 1996). Like nitrogen, the availability of phosphorus is a limiting factor in plant growth. The most common form of application for plants is as phosphate. The major difference from nitrates is that phosphate is not nearly as soluble. Consequently phosphate is normally applied as a solid fertiliser, and less frequently than nitrate. In river systems the main source of dissolved phosphate is from detergents and soaps that come through sewage treatment works. Sewage treatment works remove very little of the phosphate from detergents present in waste water, except where specific phosphate-stripping units are used. The largest amount of phosphate in river systems is normally attached to particles of sediment. Rodda *et al.* (1999) report maximum dissolved reactive phosphorus levels of 0.2 mg/l but total phosphorus levels of 1.6 mg/l. This is for intensive dairy production, where the majority of phosphate is from agricultural fertilisers.

Phosphates are a major contributor to eutrophication problems. The fact that they are bound to sediments means that they often stay in a river system for a long period of time. Improvements in water quality for a river can often be delayed substantially by the steady release of phosphate from sediments on the river bed.

Chlorine

Chlorine is not normally found in river water. It is used as a disinfectant in the supply of drinking water. It is used because it is toxic to bacteria and relatively short lived. More common to find in river water samples is the chloride ion. This may be an indicator of sewage pollution as there is a high chloride content in urine. Chloride ions give the brackish taste of sea water, the threshold for taste being around 300 mg/l. The European Commission limit for drinking water is 200 mg/l.

Heavy metals

'Heavy metals' is the term applied to metals with an atomic weight greater than 6. They are generally only found in very low levels dissolved in fresh water, but may be found in bed load sediments. In acidic waters metals can be dissolved (i.e. found in ionic form). They are often toxic in concentrations above trace levels. The toxicity, in decreasing order, is mercury, cadmium, copper, zinc, nickel, lead, chromium, aluminium and cobalt (Gray, 1999). The sources of heavy metals in the aquatic environment are almost always industrial. Sewage sludge (the product of sedimentation at a sewage treatment works) is frequently heavy metal-rich, derived from industry discharging waste into the sewerage system. It is only when untreated sewage is discharged into a river that the heavy metals can be found. Where there is a combined sewage and storm-water drainage system for an urban area, untreated sewage can be discharged during a storm event when the sewage treatment works cannot cope with the extra storm water.

WATER-QUALITY MEASUREMENT

The techniques used for water-quality analysis vary considerably depending on equipment available and the accuracy of measurement required. For the highest accuracy of measurement water samples should be taken back to a laboratory, but this is not always feasible. There are methods that can be carried out in the field to gain a rapid assessment of water quality. Both field and laboratory techniques are discussed on the following pages. Before discussing the measurement techniques it is important to consider how to sample for water quality.

Sampling methodology

It is difficult to be specific on how frequently a water sample should be taken, or how many samples represent a given stretch of water. The best way of finding this out is to take as many measurements as possible in a trial run. Then statistical analyses can be carried out to see how much difference it would have made to have had fewer measurements. By working backwards from a large data set it is possible to deduce how few measurements can be taken while still maintaining some accuracy of over-all assessment. An example of this type of approach, when used for the reduction in a hydrometry network, is in Pearson (1998). The main concern is that there are enough measurements to capture the temporal variability present and that the sample site is adequately representative of your river stretch.

One important consideration that needs to be understood is that the sample of water taken at a particular site is representative of all the catchment above it, not just the land use immediately adjacent. Adjacent land use may have some influence on the water quality of a sample, but this will be in addition to any affect from land uses further upstream which may be more significant.

Gravimetric methods

Gravimetric analysis depends on the weighing of solids obtained from a sample by evaporation, filtration or precipitation (or a combination of these three). This requires an extremely accurate weighing balance and a drying oven, hence it is a laboratory technique rather than a field one. An example of gravimetric analysis is the standard method for measuring total dissolved solids (TDS). This is to filter a known volume of water through 0.45 μm

(1 micron = one-millionth of a metre or one-thousandth of a millimetre) filter paper. The sample of water is then dried at 105 °C and the weight of residue left is the TDS.

Other examples of gravimetric analysis are total suspended solids and sulphates (causing a precipitate and then weighing it).

Volumetric methods

Volumetric analysis is using titration techniques to find concentrations of designated substances. It is dependent on measuring the volume of a liquid reagent (of known concentration) that causes a visible chemical reaction. This is another laboratory technique as it requires accurate measurements of volume using pipettes and burettes. Examples of this technique are chloride and dissolved oxygen (using the Winkler method).

Colorimetry

Colorimetric analysis depends on a reagent causing a colour to be formed when reacting with the particular ion you are interested in measuring. The strength of colour produced is assumed to be proportional to the concentration of the ion being measured (Beer's law). The strength of colour can then be assessed using one of four techniques: comparison tubes, colour discs, colorimeter, or spectrophotometer.

Comparison tubes are prepared by using standard solutions of the ion under investigation which the reagent is added to. By having a range of standard solutions the strength of colour can be compared (by eye) to find the concentration of the water sample. The standard solutions will fade with time and need remaking, hence this is a time-consuming method.

Colour discs use the same principle as comparison tubes, except in this case the standards are in the form of coloured glass or plastic filters. The coloured sample is visually compared to the coloured disc to find the corresponding concentration. It is possible to buy colour disc kits that come with small packets of reagent powder for assessment of a particular ion.

This method is extremely convenient for rapid field assessment, but is subjective and prone to inaccuracy.

A colorimeter (sometimes called an absorptiometer) takes the subjective element out of the assessment. It is similar to a turbidity meter in that a beam of light is shone through the reagent in a test tube. The amount of light emerging from the other side is detected by a photo-electric cell. The darker the solution (caused by a high concentration of reactive ion) the less light emerges. This reading can then be compared against calibrations done for standard solutions.

A spectrophotometer is the most sophisticated form of colorometric assessment. In this case instead of a beam of white light being shone through the sample (as for the colorimeter) a specific wavelength of light is chosen. The wavelength chosen will depend on the colour generated by the reagent and is specified by the reagent's manufacturer.

There are a range of spectrophotometers available to perform rapid analysis of water quality in either a laboratory or field situation. Many ions of interest in water-quality analysis can be assessed using colorimetric analysis. These include nitrate, nitrite, ammonia and phosphate.

Ion-selective electrodes

In a similar vein to pH meters ion-selective electrodes detect particular ions in solution and measure the electrical potential produced between two reactive substances. The tip of the electrode in the instrument has to be coated with a substance that reacts with the selected ion. With time the reactive ability of the electrode will decrease and need to be replaced. Although convenient for field usage and accurate, the constant need for replacing electrodes makes these an expensive item to maintain. There are ion-selective electrodes available to measure dissolved oxygen, ammonium, nitrate, calcium, chloride and others.

Spectral techniques

When ions are energised by passing electricity through them, or in a flame, they produce distinctive colours. For instance, sodium produces a distinctive yellow colour, as evidenced by sodium lamps used in some cars and street lamps. Using spectral analysis techniques the light intensity of particular ions in a flame are measured and compared to the light intensity from known standard solutions. The most common form of this analysis is atomic absorption spectrophotometry, a laboratory technique which is mostly used for metallic ions.

PROXY MEASURES OF WATER QUALITY

Any measurement of water quality using individual parameters is vulnerable to the accusation that it represents one particular point of time but not the overall water quality. It is often more sensible to try and assess water quality through indirect measurement of something else that we know is influenced by water quality. Two such proxy measures of water quality are provided by biological indicators and analysis of sediments in the river.

Biological indicators

Aquatic fauna normally remain within a stretch of water and have to try and tolerate whatever water pollution may be present. Consequently the health of aquatic fauna gives a very good indication of the water quality through a reasonable period of time. There are two different ways that this can be done: catching fauna and assessing their health; or looking for the presence and absence of key indicator species.

Fish surveys are a common method used for assessing the overall water quality in a river. It is an expensive field technique as it requires substantial human resources: people to wade through the water with electric stun guns and then weigh and measure stunned fish. When this is done regularly it gives very good background information on the overall water quality of a river.

More common are biological surveys using indicator species, particularly of macro-invertebrates.

Kick sampling uses this technique. A bottom-based net is kicked into sediment to catch any bottom-based macro-invertebrates, which are then counted and identified. There are numerous methods that can be used to collate this species information. In Britain the BMWP (Biological Monitoring Working Party) score is commonly used and provides good results. Species are given a score ranging from 1 to 10; with 10 representing species that are extremely intolerant to pollution. The presence of any species is scored (it is purely presence/absence, not the total number) and the total for the kick sample calculated. The BMWP score has a maximum of 250. Other indicator species scores include the Chandler index and the ASPT (Average Score Per Taxon). Details of these can be found in a more detailed water-quality assessment text such as Chapman (1996).

Another example of an indicator species used for water-quality testing is *Escherichia coli* (*E. coli*). These are used to indicate the presence or absence of faecal contamination in water. *E. coli* is a bacteria present in the intestines of all mammals and excreted in large numbers in faeces. Although one particular strain (*E. coli*$_{157}$) has toxic side effects the vast majority of *E. coli* are harmless to humans. Their presence in a water sample is indicative of faecal pollution, which may be dangerous because of other pathogens carried in the contaminated water. They are used as an indicator species because they are easy to detect, while viruses and other pathogens are extremely difficult to measure. Coliform bacteria (i.e. bacteria of the intestine) are detected by their ability to ferment lactose, producing acid and gas (Tebbutt, 1993). There are specific tests to grow *E. coli* in a lactose medium, which allow the tester to derive most probable number per 100 ml (MPN/100 ml).

Sediments

The water in a channel is not the only part of a river that may be affected by water pollution. There are many substances that can build up in the sediments at the bottom of a river and provide a record of pollution. There are two big advantages to this method for investigating water quality: the sediments will reflect both instantaneous large pollution events and long, slow contamination at low levels; and if the river is particularly calm in a certain location the sediment provides a record of pollution with time (i.e. depth equals time). Not all water pollutants will stay in sediments, but some are particularly well suited to study in this manner (e.g. heavy metals and phosphorus).

The interpretation of results is made difficult by the mobility of some pollutants within sediments. Some metals will bind very strongly to clay particles in the sediments (e.g. lead and copper), and you can be fairly certain that their position is indicative of where they were deposited. Others will readily disassociate from the particles and move around in the interstitial water (e.g. zinc and cadmium) (Alloway and Ayres, 1997). In this case you cannot be sure that a particularly high reading at one depth is from deposition at any particular time.

MODELLING WATER QUALITY

The numerical modelling of water quality is frequently required, particularly to investigate the effects of particular water-quality scenarios. The type of problems investigated by modelling are: the impact of certain levels of waste discharge on a river (particularly under low flow levels); recovery of a water body after a pollution event; the role of backwaters for concentration of pollutants in a river; and many more. The simplest water-quality models look at the concentration of a certain pollutant in a river given knowledge about flow conditions and decay rates of the pollutant. The degradation of a pollutant with time can be simulated as a simple exponential decay rate equation. A simple mass balance approach can then be used to calculate the amount of pollutant left in the river after a given period of time (James, 1993). More complex models build on this approach and incorporate ideas of diffusion, critical loads of pollutants and chemical reaction between pollutants in a river system. If the problem being researched is to track pollutants down a river then it is necessary to incorporate two- or three-dimensional representation of flow hydraulics. There are numerous water-quality models available in the research literature, as well as those used by consultants and water managers.

EUTROPHICATION

'Eutrophication' is the term used to describe the addition of nutrients to an aquatic ecosystem that leads to an increase in net primary productivity. The term comes from limnology (the study of lakes and freshwater ponds) and is part of an overall classification system for the nutrition, or trophic, level of a freshwater body. The general classification moves from oligotrophic (literally 'few nutrients'), to eutrophic ('good nutrition') and ends with hypertrophic ('excess nutrients'). In limnology this classification is viewed as part of a natural progression for bodies of water as they fill up with sediment and plant matter. Eutrophication is a natural process (as part of the nitrogen and phosphorus cycles), but it is the addition of extra nutrients from anthropogenic activity that attracts the main concern in hydrology. In order to distinguish between natural and human-induced processes the term cultural eutrophication is sometimes used to identify the latter.

The major nutrients that restrict the extent of a plant's growth are potassium (K), nitrogen (N) and phosphorus (P). If you buy common fertiliser for a garden you will normally see the K:N:P ratio expressed to indicate the strength of the fertiliser. For both aquatic and terrestrial plants nitrogen is required for the production of chlorophyll and green leaves, while potassium and phosphorus are needed for root and stem growth. In the presence of abundant nitrogen and phosphorus (common water pollutants, see pp. 116–18), aquatic plant growth, including algae, will increase dramatically. This can be seen as positive as it is one way of removing the nitrate and phosphate from the water, but overall it has a negative impact on the river system. The main negative effect is a depletion of dissolved oxygen caused by bacteria decomposing dead vegetative matter in the river. In temperate regions this is a particular problem in the autumn when the aquatic vegetation naturally dies back. In tropical regions it is a continual problem. A second negative effect is from algal blooms. In 1989 there was an explosion in cyano-bacteria numbers in Rutland Water, a reservoir supplying drinking water in central England (Howard, 1994). (NB These are also called blue-green algae, despite being a species of bacteria.) The cyano-bacteria produce toxins as waste products of respiration that can severely affect water quality. In the 1989 outbreak several dogs and sheep that drunk water from Rutland Water were poisoned, although no humans were affected (Howard, 1994). In an effort to eliminate future problems the nutrient-rich source water for Rutland Water is supplemented with water from purer river water pumped from further afield.

Table 8.4 shows some of the indicators used in a quantitative example of defined trophic levels developed for the Organisation for Economic Co-operation and Development (OECD). The chlorophyll is an indicator of algal growth in the water, while phosphorus and dissolved oxygen are more traditional water-quality measures. The dissolved oxygen is taken from the bottom of the lake because this is where the vegetative decomposition is taking place. The dissolved oxygen level near to the surface will vary more because of the proximity to the water/air interface and the oxygen produced in photosynthesis by aquatic plants. It is worth noting that heavily eutrophied water samples will sometimes have a dissolved

oxygen greater than 100 per cent. This is due to the oxygen being produced by algae which can supersaturate the water.

CONTROLLING WATER QUALITY

Waste water treatment

The treatment of waste water is a relatively simple process that mimics natural processes in a controlled, unnatural environment. The treatment processes used for industrial waste water is dependent on the type of waste being produced. In this section the processes described are those generally found in sewage treatment rather than in specialised industrial waste water treatment.

There are two major objectives for successful sewage treatment: to control the spread of disease from waste products and to break down the organic waste products into relatively harmless metabolites (i.e. by-products of metabolism by bacteria etc.). The first objective is achieved by isolating the waste away from animal hosts so that viruses and other pathogens die. The second objective is particularly important for the protection of where the treated effluent ends up – frequently a river environment.

In Britain the first attempt to give guidelines for standards of sewage effluent discharge were provided by the Royal Commission on Sewage Disposal which sat between 1898 and 1915. The guidelines are based on two water-quality parameters described earlier in this chapter: suspended solids and biochemical

Table 8.4 OECD classification of lakes and reservoirs for temperate climates

Trophic level	Average total P (mg/l)	Dissolved oxygen (% saturation)	Max. chlorophyll (mg/l) (at depth)
Ultra-oligotrophic	0.004	>90	0.0025
Oligotrophic	0.01	>80	0.008
Mesotrophic	0.01–0.035	40–89	0.008–0.025
Eutrophic	0.035–0.1	0–40	0.025–0.075
Hypertrophic	>0.1	0–10	>0.075

Source: Adapted from Meybeck *et al.* (1989)

oxygen demand (BOD). The Royal Commission set the so-called 30:20 standard which is still applicable today (i.e. 30 mg/l of suspended solids and 20 mg/l of BOD). The standard was based on a dilution ratio of 8:1 with river water. Where river flow is greater than eight times the amount of sewage effluent discharge the effluent should have a TSS of less than 30 mg/l and a BOD of less than 20 mg/l. There was also the recommendation that if the river is used for drinking water extraction further downstream the standard should be tightened to 10:10. This was used as a recommendation until the 1970s when a system of legal consents to discharge were introduced (see p. 125).

The processes operating at a waste water treatment works are very simple. They are summarised below and in Figure 8.6 (NB Not every sewage treatment works will have all of these processes present.)

1   Primary treatment: screening and initial settlement.
2   Secondary treatment: encouraging the biological breakdown of waste and settling out of remaining solids. This can take place either in trickle bed filters or activated sludge tanks. The main requirement is plenty of oxygen to allow micro-organisms to break down the concentrated effluent.
3   Tertiary treatment: biodigestion of sludge (from earlier settling treatment); extra treatment of discharging effluent to meet water-quality standards (e.g. phosphate stripping, nitrate reduction).

Raw sewage entering a sewage treatment works is approximately 99.9 per cent water (Gray, 1999). This is derived from water used in washing and toilet flushing, and also from storm runoff in an urban environment where there is a combined sewage/stormwater drainage scheme. Of the solids involved, the majority are organic and about half are dissolved in the water (TDS). Of the organic compounds the breakdown is approximately 65 per cent nitrogenous (proteins and urea), 25 per cent carbohydrates (sugars, starches, cellulose) and 10 per cent fats (cooking oils, grease, soaps) (Gray, 1999). Typical values for TSS and BOD at different stages of sewage treatment are provided in Table 8.5.

In tertiary treatment an effort is sometimes (but not always) made to reduce the level of nitrate and phosphorus in the discharged waste. In some cases

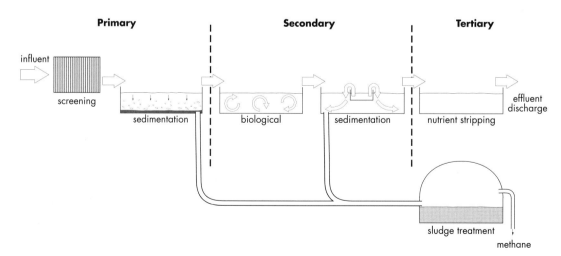

Figure 8.6 Schematic representation of waste water treatment from primary through to tertiary treatment, and discharge of the liquid effluent into a river, lake or the sea.

Table 8.5 Changes in suspended solids and biochemical oxygen demand through sewage treatment. These are typical values which will vary considerably between treatment works

Stage of treatment	Suspended solids (mg/l)	BOD (mg/l)
Raw sewage	400	300
After primary treatment	150	200
After biological treatment	300	20
Effluent discharged to river	30	20

this is achieved through final settling ponds where the growth of aquatic flora is encouraged and the nutrients are taken up by the plants before discharge into a stream. Of particular use are reeds which do not die back during the winter (in temperate regions). This is a re-creation of natural wetlands that have been shown to be extremely efficient removers of both nitrogen and phosphorus from streams (e.g. Russel and Maltby, 1995). Other methods of phosphate removal are to add a lime or metallic salt coagulant that causes a chemical reaction with the dissolved phosphorus so that an insoluble form of phosphate settles out. This is particularly useful where the receiving water for the final effluent has problems with eutrophication. The average phosphorus concentration in raw sewage is 5–20 mg/l, of which only 1–2 mg/l is removed in biological treatment.

In some cases, particularly in the USA, chlorination of the discharging effluent can take place. Chlorine is used as a disinfectant to kill any pathogens left after sewage treatment. This is a noble aim but creates its own difficulties. The chlorine can attach to organic matter left in the effluent and create far worse substances such as polychlorinated biphenyl (PCB) compounds. Another safer form of disinfection is to use ultraviolet light, although this can be expensive to install and maintain.

Source control

The best way of controlling any pollution is to try and prevent it happening in the first place. In order to achieve this the differentiation has to be recognised between point source and diffuse pollutants (see p. 110). When control over the source of pollutants is achieved dramatic improvements in river-water quality can be achieved. An example of this is shown in the Case Study of the Nashua river in Massachusetts, USA.

Controlling point source pollutants

The control of point source pollutants cannot always be achieved by removing that point source. It is part of water resource management to recognise that their may be valid reasons for disposing of waste in a river; the trick is to ensure that that disposal creates no harmful side effects. In the United Kingdom the control of point source pollution is through discharge consents. These provide a legal limit for worst-case scenarios – for example, at individual sewage treatment works they are usually set with respect to TSS, BOD and ammonia (sometimes heavy metals are included), and calculated to allow for low flow levels in the receiving stream (see the technique box for calculating discharge consents on p. 127). There is also an obligation to comply with the European Union Urban Waste Water directive.

Controlling diffuse source pollutants

The control of diffuse source water pollution is much harder to achieve. In an urban environment this can be achieved through the collection of stormwater drainage into artificial wetlands where natural processes can lessen the impact of the pollutants on the draining stream. Of particular

Case study

CONTROLLING WATER QUALITY OF THE NASHUA RIVER

The Nashua river is an aquatic ecosystem that has undergone remarkable change in the last one hundred years. It drains an area of approximately 1,400 km² in the state of Massachusetts, USA, and is a tributary of the much larger Merrimack river which eventually flows into the sea in Boston Harbor (see Figure 8.7). The land use of the Nashua catchment is predominantly forest and agricultural, with a series of towns along the river. It is the industry associated with these towns that has brought about the changes in the Nashua, predominantly through the twentieth century. The latter-day changes are well illustrated by the two photographs at the same stretch of the Nashua, in 1965 and 1995 (see Plates 9 and 10).

Prior to European colonisation of North America the Nashua valley was home to the Nashaway tribe, and the Nashua river could be considered to be in a pristine condition. With the arrival of European settlers to New England the area was used for agriculture and the saw milling of the extensive forests. The Industrial Revolution of the nineteenth century brought manufacturing to the area and mills sprang up along the river. By the middle of the twentieth century the small towns along the Nashua (Gardner, Fitchburg, Leominster and Nashua) were home to paper, textile and shoe factories, many of which were extracting water from the river and then discharging untreated waste back into the river. The photograph of the Nashua in 1965 (Plate 9) is indicative of the pollution problems experienced in the river; in this case dye from a local paper factory has turned the river red. Under the US water-quality classification scheme the river was classified as U: unfit to receive sewage.

In 1965 the Nashua River Clean-Up Committee was set up to try to instigate a plan of restoring the water quality in the river. This committee later became the Nashua River Watershed Association (NRWA) which still works today to improve water-quality standards in the area. Between 1972 and 1991 eleven waste water-treatment plants were constructed or upgraded to treat waste from domestic, and to a lesser extent from industrial, sources in the catchment. These were built using grants from the state and federal government as part of a strategy to improve the river from U to B status (fit for fishing and swimming). Through this control of point source pollution the river-water quality has improved dramatically as can be seen in the second photograph of the river (Plate 10). The river has attained B status and is an important recreational asset for the region. It has not returned to a pristine state, though, and is unlikely to while there is still a significant urban

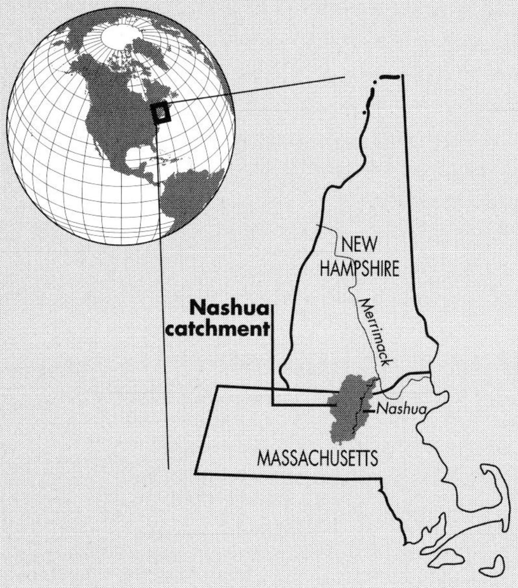

Figure 8.7 Location of the Nashua catchment in north-east USA.

population in the catchment. There are problems with combined sewage and stormwater drainage systems discharging untreated waste into the river during large storms, and also diffuse pollution sources – particularly in the urban environment. However, during the latter half of the twentieth century the Nashua river has had its water quality transformed from an abiotic sewer into a clean river capable of maintaining a healthy salmonid fish population. This has largely been achieved through the control of point pollution sources.

The author gratefully acknowledges the Nashua River Watershed Association for supplying much of this information and Plates 9 and 10. For more information the NRWA can be contacted at nrwa@ma.ultranet.com

Technique: calculating discharge consents

In England and Wales the setting of discharge consents for point source pollution control is carried out by the Environment Agency. A discharge consent gives a company the right to dispose of a certain amount of liquid waste into a river system so long as the pollution levels within the discharge are below certain levels. To calculate what those critical levels are a series of computer programs are used. These computer programs are in the public domain and can be obtained from the Environment Agency. They use very simple principles that are described here.

The main part of the discharge consent calculation concentrates on a simple mass balance equation:

$$C_D = \frac{Q_U C_U + Q_E C_E}{Q_U + Q_E}$$

where Q refers to the amount of flow (m^3/s) and C the concentration of pollutant. For the subscripts: D is for downstream; U is for upstream (i.e. the background); and E is for the effluent.

With this mass balance equation the downstream concentration can be calculated with varying flows and levels of effluent concentrations. This variation in flow and concentration is achieved through a computer program running a Monte Carlo simulation.

In this case the Monte Carlo simulation involves a random series of values for Q_U, Q_E, C_U and C_E drawn from an assumed distribution for each variable. It is assumed that the distributions are log-normal in shape (see Figure 8.8); therefore by using the data in Table 8.6 the actual distribution for each variable is simulated. Once the distribution for each variable is known then a random variable is chosen from that distribution. In the case of a log-normal distribution this means that it is most likely to be close to the mean value but more likely to be above than below the mean (see Figure 8.8). In a Monte Carlo simulation the value of C_D is calculated many times (often set to 1,000) so that a distribution for C_D can be drawn. The consent to discharge figure is taken from the distribution of C_D, usually looking at the 90

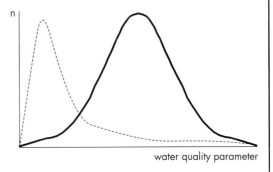

Figure 8.8 A log-normal distribution (dashed line) compared to a normal distribution (solid line).

Table 8.6 Parameters required to run a Monte Carlo simulation to assess a discharge consent

Variable	Required data
River flow (Q_U)	Mean daily flow and Q_{95}
Upstream river quality (C_U)	Mean value and standard deviation
Effluent flow (Q_E)	Mean value and standard deviation
Effluent quality (C_E)	Mean and standard deviation

or 95 percentile values (i.e. the target will be achieved 90 or 95 per cent of the time).

The variables that are required for calculating a consent to discharge (see Table 8.6) are derived from normal hydrological data. River flow data can be derived from a flow duration curve (see Chapter 7). The water-quality information requires at least 3–4 years of regular measurements. The values to describe Q_E and C_E will either be known or are to be varied in the simulation in order to derive a consent to discharge value.

In short, the person calculating the discharge consent inserts values from Table 8.6 into the Monte Carlo simulation. This will then produce the 95 percentile value of C_D. If that value is too high (i.e. too much pollution) then the simulation is run again using lower values for C_E until a reasonable value is derived. Once the reasonable value has been reached then the 95 percentile value of C_E is taken as the discharge consent. The definition of a 'reasonable value' will be dependent on the designated use of the river. Rivers with high-class fisheries and those with abstraction for potable supply have much higher standards than for other uses.

The approach described here can be used to calculate a consent to discharge for such water-quality parameters as BOD and TSS. When a calculation is being carried out for ammonia then more data are required to describe the water quality in the receiving river. Parameters such as pH, temperature, alkalinity, TDS and dissolved oxygen (all described with mean and standard deviation values) are required so that chemical reaction rates within the river can be calculated.

A scheme such as discharge consents provides a legal framework for the control of point source pollutants, but the actual control comes about through implementing improved waste water treatment.

concern is runoff derived from road surfaces where many pollutants are present as waste products from vehicles. Hamilton and Harrison (1991) suggest that although roads only make up 5–8 per cent of an urban catchment area they can contribute up to 50 per cent of the TSS, 50 per cent of the total hydrocarbons and 75 per cent of the total heavy metals input into a stream. The highest pollutant loading comes during long, dry periods which may be broken by flushes of high rainfall (e.g. summer months in temperate regions). In this case the majority of pollutants reach the stream in the first flush of runoff. If this runoff can be captured and held then the impact of these diffuse pollutants is lessened. This is common practice for motorway runoff where it drains into a holding pond before moving into a nearby water course.

Another management tool for control of diffuse pollutants is to place restrictions on land management practices. An example of this is in areas of England that have been designated either a Nitrate Vulnerable Zone (NVZ) or a Nitrate Sensitive Area (NSA), predominantly through fears of nitrate contamination in aquifers. In NSAs the agricultural practices of muck spreading and fertilising with nitrates are heavily restricted. This type of control relies on tight implementation of land use planning – something that is not found uniformly between countries, or even within countries.

SUMMARY

The measurement and management of water quality in a river is an important task within hydrology. To carry this out, a knowledge of the pollution type, pollution source (assuming it is not natural) and pathways leading into the stream are important. Equally, it is important to know the flow regime of any receiving river so that dilution rates can be assessed. There are methods available to control water quality, whether through treatment at point sources (e.g. waste water treatment) or control throughout a catchment using land use planning.

FURTHER READING

Chapman, D. (ed.) (1996) *Water quality assessments: a guide to the use of biota, sediments and water in environmental monitoring* (2nd edn). Chapman & Hall, London.
A comprehensive guide to water-quality assessment.

Gray, N.F. (1999) *Water technology: an introduction for environmental scientists and engineers*. Arnold, London.
An introduction to the engineering approach for controlling water pollution.

ESSAY QUESTIONS

1 Explain the Hjulstrom curve and describe its importance for suspended sediment loading in a river.

2 Discuss the importance of the BOD_5 test in the assessment of overall water quality for a river.

3 Compare and contrast the direct measurement of water-quality parameters to the use of proxy measures for the overall assessment of water quality in a river.

4 Explain the major causes of enhanced (or cultural) eutrophication in a river system and describe the measures that may be taken to prevent it occurring.

5 Describe the different techniques that can be used to lessen the impact of pollution from agricultural sources in a river catchment.

9

HYDROLOGY IN A CHANGING WORLD

LEARNING OBJECTIVES

When you have finished reading this chapter you should have:

- An understanding of the main issues of change that affect hydrology.
- An understanding of how hydrological investigations are carried out to look at issues of change.
- A knowledge of the research literature and main findings in the issues of change.
- A knowledge of case studies looking at change in different regions of the world.

We live in a world that is constantly adjusting to change. This applies from the natural, through to the economic world and is fundamental to the way that we live our lives. The theory of evolution proposes that in order to survive each species on the planet is changing over a long time period (through natural selection) in order to adapt to its ecosystem fully. Equally, economists would say that people and businesses need to adapt and change to stay competitive in a global economy. If, as was argued in the introductory chapter of this book, water is fundamental to all elements of our life on this planet, then we would expect to see hydrology constantly changing to keep up with our changing world. It is perhaps no great surprise to say that hydrology has, and is, changing – but not in all

areas. The principle of uniformitarianism states in its most elegant form: 'the present is the key to the past'. Equally, it could be said that the present is the key to the future and we can recognise this with respect to the fundamentals of hydrology. By the end of the twenty-first century people may be living in a different climate from now, their economic lives may be unlike ours, and almost certainly their knowledge of hydrological processes will be greater. However, the hydrological processes will still be operating in the same manner, although it may be at differing rates than those that we measure today.

The early chapters of this book have been concerned with hydrological processes and our assessment of them. Our knowledge of the processes will improve, and our methods of measurement and estimation will get better, but the fundamentals will still be the same. In this final chapter several hydrological issues are explored with respect to change. They are, by and large, management problems: how we respond to changes in patterns of consumption; increasing population pressure and possible changes in climate. The topics discussed here are not exhaustive in covering all issues of change that might be expected in the near future, but they do reflect some of the major concerns. It is meant as an introduction to issues of change and how they affect hydrology; other books cover some of these issues in far more depth (e.g. McDonald and Kay, 1988; Acreman, 2000). The first broad topic of discussion is water resource management, particularly at the local scale. The second topic is the one that dominates the research literature in natural sciences at present: climate change. The third and fourth topics are concerned with the way we treat our environment and the effect this has on water resources: land use change and groundwater depletion. The final topic is in urban hydrology – of great concern, with an ever-increasing urban population all around the world.

WATER RESOURCE MANAGEMENT

When the topic of water resource management is discussed it is often difficult to pinpoint exactly what authors mean by the term. Is it concerned with all aspects of the hydrological cycle or only with those of direct concern to humans, particularly water consumption? As soon as the term 'resource' is introduced then it automatically implies a human dimension. Water is a resource because we need it, and there are ways that we can manipulate its provision, therefore water resource management is a very real proposition. If we are going to manage the water environment is it purely for consumption or are there other uses that need protection and management? During the twentieth century there has been a large-scale rise in the amount of time spent at leisure. Included in this are such sports as

fishing, canoeing and boating, all of which desire clean, fully flowing rivers. Thought is now given to the **amenity value** of rivers and lakes (i.e. how useful they are as places of pleasure). Managing the water environment needs to be designed to maintain and enhance amenity values. Equally we have an obligation to protect the water environment for future generations and for other species that co-exist with the water. Therefore water resource management needs to embrace sustainable development in its good practices. It is clear that water resource management has to embrace all of these issues and at the same time adapt to changing views on what is required of water management.

Almost all of the processes found in the hydrological cycle can be manipulated in some way. Table 9.1 sets out some of these interventions and the implications of these being dealt with by those involved in water resource management. It is immediately apparent from Table 9.1 that the issues go far beyond the river boundary. For example, land use change has a huge importance for water resource management, so that any decisions on land use need to have consultation with water resource managers. It is important that a legislative framework is in place for countries so that this consultation does take place. Likewise for other areas where human intervention may have a significant impact on water resources. The issue of finding the correct management structures and legislation is investigated in the Case Study looking at how water resources have been managed in England over the past forty years (see pp. 132–134). The changing world in this case has been through increasing population pressures, but (probably more importantly) has had to adapt to changing political beliefs.

Aside from the impacts of change on hydrological processes there is also a problem in water resource management concerning the statistical techniques that we use. In a frequency analysis technique there is an inherent assumption made that a storm event with similar antecedent conditions, at any time in the streamflow record, will cause the same size of storm. We assume that the hydrological regime is stationary with time. Under conditions of land use

Table 9.1 Manipulation of hydrological processes of concern to water resource management

Hydrological process	Human intervention	Impact
Precipitation	Cloud seeding	Increase rainfall (?)
Evaporation	Irrigation	Increase evaporation rates
	Change vegetation cover	Alter transpiration and interception rates
	Change rural to urban	Increase evaporation rates
Storage	Change land use	Alter infiltration rates
	Aquifer storage and recovery (ASR)	Manipulating groundwater storage
	Land drainage	Lowering of local water tables
	Building reservoirs	Increasing storage
Runoff	Change land use	Alter overland flow rates
	Land drainage	Rapid runoff
	River transfer schemes	Alter river flow rates
	Water abstraction	Removing river water and groundwater for human consumption

---

Case study

**CHANGING STRUCTURES OF WATER RESOURCE MANAGEMENT –
ENGLAND AS AN EXAMPLE**

The major issues of concern for water resource management in England are water supply, waste disposal, pollution and water quality, and fisheries/aquatic ecosystems management. Other inter-related issues that come into water resource management are flood defence and navigation. Historically it is the first three that have dominated the political agenda in setting up structures to carry out water resource management in England.

History of change

Towards the end of the nineteenth century great municipal pride was taken in the building of reservoirs to supply water to urban centres in England. At the same time many sewage treatment works were built to treat waste. These were built and run by local councils and replaced a previously haphazard system of private water supply and casual disposal of waste. At this stage water resource management resided firmly at the local council level. This system continued until the Water Act of 1973 was passed: a bill that caused a major shake up of water resource management in England and Wales. The major aim of the 1973 Act was to introduce holistic water management through administrative boundaries that were governed by river catchments rather than political districts. There was some success in this regard, with Regional Water Authorities (RWAs) taking over the water management issues listed above from local councils and other bodies. One of the difficulties with this management structure was the so-called 'poacher–gamekeeper' problem whereby the RWAs were in charge of both waste disposal and pollution control, creating a conflict of interest in water management. Throughout their existence the RWAs operated from a diminishing funding base which led to a lack of investment in waste treatment facilities. It was obvious by the end of the 1980s that a raft of upcoming European Community legislation on water quality would require a huge investment in waste treatment to meet water-quality standards. The government at the time decided that this investment was best supplied through the private sector and in 1989 a new Water Act was introduced to privatise the supply of drinking water and waste water treatment. This has created a set of private water companies with geographic boundaries essentially the same as the RWAs. At the same time a new body, the National Rivers Authority (NRA), was set up to act as a watchdog for water quality. This management structure is still in place in 2002, except that since 1996 the NRA has been subsumed within a larger body: the Environment Agency. The Environment Agency (amongst other duties) monitors river water quality, prosecutes polluters, issues licences for water abstraction and treated waste disposal.

How has this change affected water resource management?

The answer to this question can be answered by looking at figures for water abstraction and measured water quality over time.

Figure 9.1 shows the water abstracted for supply in England and Wales from 1961 to 2000. During the period of public control (whether councils or RWAs) there was a steady rise in the amount of water abstracted, apart from a blip in the mid-1970s when there were two particularly dry years. Since privatisation in 1989 there has been a flattening and then decline in the amount of water abstracted. This decline cannot be accounted for by the population, which has shown a gradual increase during the same time period (see Figure 9.1). There are two causes of this decline: less water being used for industry due to a decline in the industrial sector (although this has been in decline since the early 1980s), and a drop in the amount of leakage from the supply network. This second factor has been forced upon the water

companies by political pressure, particularly following a drought in 1995 and allegations of water supply mismanagement in Yorkshire Water plc. The reduction in leakage has required considerable investment of capital into water supply infrastructure. Overall the water abstracted for public supply is now at the same level as the late 1970s, despite a population rise of nearly 4 million in the corresponding period. The decline has also been achieved despite an increase in the amount of water consumed per household. The United Kingdom has the highest water consumption per capita in Europe and this is rising, a reflection of changing washing habits and an increase in the use of dishwashers. This decline in water abstractions is good for the aquatic environment as it allows a more natural river regime and groundwater system to operate with less human intervention.

It is more difficult to ascertain how the changing management structure has affected water quality in England and Wales because the ways of describing water quality have changed with time. Figure 9.2 shows river-water quality assessment using three different scales. The figures shown are achieved by sampling water quality (over a period of time, normally years) for hundreds of river reaches around the country. During

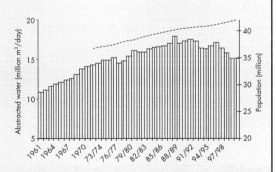

Figure 9.1 Abstracted water for England and Wales 1961–2000 (bar chart), with population for England 1971–2000 shown as a dashed line.
Source: Data from OFWAT and various other sources

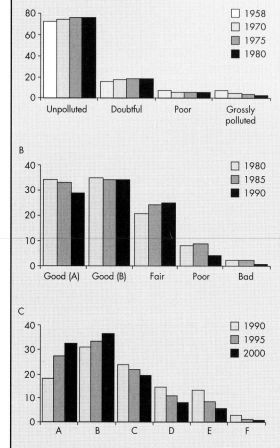

Figure 9.2 Water-quality assessment for three periods between 1958 and 2000. Explanation of the differing scales is given in the text.

water quality (only the chemical results are shown in Figure 9.2). The measurements for 1990 have been recalculated onto the five-point scale to provide some continuity between assessments. Since 1990 there has been a rise in the two highest categories of water quality at the expense of all the others, a response to the extra investment in waste water treatment provided by the privatised water companies. The percentage figures for 2000 suggest that the lowest category of water quality has been almost eliminated, although when this is recalculated as river length it shows that there are approximately 162 km of extremely poor-quality river reaches (out of a total 40,588 km assessed).

In summary it has to be said that the biggest impact on the water environment for England and Wales has been the privatisation of supply and waste water treatment, and the setting up of a separate environmental watchdog organisation, in 1989. Prior to this the water quality remained static or worsened between local authority control and the RWAs and the amount of water abstracted continued to rise. Although the integration of water management into a holistic structure based around the water catchments (i.e. the RWAs) was a noble idea it made very little difference to the crude measures shown here. Since privatisation the water quality has improved and the total amount of water abstracted has decreased. Fundamentally, the reason for this is that the investment in infrastructure has risen dramatically since privatisation. It is probably reasonable to surmise that given the same increase in investment a RWA structure would have seen a similar improvement. It was never likely, in the political climate of the late 1980s, that this investment would have come from the public purse. It has been left to private companies, and more particularly their customers, to pick up the cost of that investment. The example of changing water management structure in England and Wales shows us that this type of change can have significant impacts on the overall hydrology of a region.

the first period, when the control was either local council or RWAs, there was very little change when assessed on a four-point scale. After 1980 the scale was changed to five points, and during this period of predominantly RWA control the A-grade water quality declined while the percentage of 'fair' water quality river reaches increased. In 1995 the scale was changed again to make it five points, and also a differentiation was made between biological and chemical monitoring of or climate change it is quite possible that these conditions will not be met. This makes it difficult to put much faith in a technique such as frequency analysis when it is known that the hydrological regime has changed during the period of record.

The way that water resource management copes with change will be one of the great issues for debate and research in the coming century.

CLIMATE CHANGE

At the start of the twenty-first century climate change is the biggest environmental talking point, dominating the scientific media and research agenda. Any unusual weather patterns are linked to the greenhouse effect and its enhancement by humans. The winter of 2000–2001 in England was one of the wettest on record and there was widespread flooding. At the same time the east coast of the South Island of New Zealand experienced one of the worst droughts on record. At various times in the media, both these events were linked to global warming. The difficulty with trying to verify any real link to climate change is that hydrological systems naturally contain a huge amount of variability. The extreme events we are experiencing now may be part of that natural variability, or they may be being pushed to further extremes by climate change. It is unlikely we will know for sure until it is too late to try and do anything about it.

Predictions from the Intergovernmental Panel on Climate Change (IPCC, 2001) suggest that the earth may experience a global surface temperature rise of 1–3.5°C over the next hundred years. Linked to this prediction are an increase in sea level of 15–95 cm and changes in the temporal and spatial patterns of precipitation. All of these predicted changes will influence the hydrological cycle in some way, but it is difficult to pinpoint exactly how. At the very simple level a temperature rise would lead to greater evaporation rates, which in turn puts more water into the atmosphere. This may lead to higher precipitation rates, or at least changes in precipitation patterns. How this impacts the

hydrology of an individual river catchment is very difficult to predict. The most common method to make predictions is to take the broad-brush predictions from a global circulation model (often at scale of 1° latitude and longitude per grid square) and downscale it to the local river catchment level. There are several methods used to downscale the data, and Wilby et al. (2000) show that the choice of method used can influence the modelling predictions dramatically.

Arora and Boer (2001) have simulated the impacts of possible future climate change on the hydrology of twenty-three major river catchments worldwide. They conclude that in warmer climates there may be a general reduction in annual mean discharge, although as some rivers showed an increase this is not absolute. For mid- to high-latitude rivers Arora and Boer (2001) concluded that there may be big changes in the timing of large runoff events that could be linked to changing seasonal times. This confirms the findings of Middelkoop et al. (2001) who predict higher winter discharges on the Rhine (Europe) from 'intensified snow-melt and increased winter precipitation'. In a similar vein Wilby and Dettinger (2000) predict higher winter flows for three river basins in the Sierra Nevada (California, USA). These higher winter flows reflect changes in the winter snowpack due to a predicted rise in both precipitation and temperature for the region. Arnell and Reynard (1996) used models of river flow to try and predict the effects of differing climate change predictions on the river flows in twenty-one river catchments in Great Britain. Their results suggest a change in the seasonality of flow and also considerable regional variation. Both these changes are by and large driven by differences in precipitation. The North West of England is predicted to become wetter while the South East becomes drier. Overall it is predicted that winters will be wetter and summers drier. This may place a great strain on the water resources for south-east England where by far the greater percentage of people live. In a more recent study Arnell and Reynard (2000) have suggested that flow duration curves are likely to become steeper, reflecting a greater variability in flow. They also predict an increase in flood magnitudes that in the case of the Thames and River Severn have 'a much greater effect than realistic land use change' (Arnell and Reynard, 2000: 21). These changes in river flow regime have important implications for water resource management in the future.

In non-temperate regions of the world predictions vary on climate change. Parry (1990) suggests rainfall in the Sahel region of Africa will stay at current levels or possibly decline by 5–10 per cent. Parry (1990) also suggests a 5–10 per cent increase in rainfall for Australia, although this may have little effect on streamflow when linked with increased evaporation from a 2°C temperature rise. Chiew *et al.* (1995) highlight the large regional variations in predictions of hydrologic change in Australia. The wet tropical regions of north-east Australia are predicted to have an increase in annual runoff by up to 25 per cent, Tasmania a 10 per cent increase, and a 35 per cent decrease in South Australia, by 2030. The uncertainty of this type of prediction is illustrated by south-east Australia where there are possible runoff changes of ±20 per cent (Chiew *et al.*, 1995). Similarly, Kaleris *et al.* (2001) attempted to model the impacts of future climate change on rivers in Greece but concluded that the 'error of the model is significantly larger than climate change impacts' and therefore no firm conclusions could be made. Overall it is difficult to make predictions for changes in hydrology as the feedback mechanisms within climate change are not properly understood.

CHANGE IN LAND USE

The implications of land use change for hydrology has been an area of intense interest to research hydrologists over the last 50 or more years. Issues of land use change affecting hydrology include increasing urbanisation (see pp. 143–145), changing vegetation cover, land drainage, and changing agricultural practices leading to salination.

Vegetation change

In Chapter 4 a Case Study showed the effect that trees have on evaporation and interception rates. This is a hydrological impact of vegetation cover change, a subject that Bosch and Hewlett (1982) review in considerable depth. In general Bosch and Hewlett conclude that the greater the amount of deforestation the larger the subsequent streamflows will be, but the actual amount is dependent on the vegetation type and precipitation amount. This is illustrated by the data in Table 9.2. In the Australian study of Crockford and Richardson (1990) the large range of values are from different size storms. The high interception losses were experienced during small rainfall events and vice versa. The interception loss from Amazonian rain forest is remarkably low, reflecting a high rainfall intensity and high humidity levels. Overall there is a high degree of variability in the amount of interception that is likely to occur. While it may be possible to say that in general a land use change that has increased tree cover will lead to a water loss, it is not easy to predict by how much that will be.

Fahey and Jackson (1997) conclude that with the loss of forest cover both low flows and peak flows increase. The low flow response is altered primarily through the increase in water infiltrating to groundwater without interception by a forest canopy. The peak flow response is a result of a generally wetter soil and a low interception loss during a storm when there is no forest canopy cover. The time to peak flow may also be affected, with a more sluggish response in a catchment with trees. In a modelling study, Davie (1996) has suggested that any changes in peak flow that result from afforestation are not gradual but highly dependent on the timing of canopy closure.

In Chapter 6 the issue of measurement scale was discussed, and it is particularly pertinent for issues of land use change. In recent years there has been considerable debate in the hydrological research literature as to how detectable the effects of deforestation are in large river catchments. Jones and Grant (1996) and Jones (2000) analysed data from a

Table 9.2 The amount of interception loss (or similar – see note below) for various canopies as detected in several studies

Canopy cover	Interception loss	Source
Eucalypt forest (Australia)	5–26% per rainfall event	Crockford and Richardson (1990)
Pine forest (Australia)	6–52% per event	Crockford and Richardson (1990)
Oak stand (Denmark)	15% of summer rainfall	Rasmussen and Rasmussen (1984)
Amazonian rain forest	9% of annual	Lloyd *et al.* (1988)
Sitka spruce (Lancashire, England)	38% of annual precipitation*	Law (1956)
Sitka spruce (Wales)	30% of annual precipitation*	Kirby *et al.* (1991)
Grassland (Wales)	18% of annual precipitation*	Kirby *et al.* (1991)
Young Douglas fir (New Zealand) (closed canopy)	27% of 7-month summer/ autumn period	Fahey *et al.* (2001)
Mature Douglas fir (NZ)	24% of 7-month summer/ autumn period	Fahey *et al.* (2001)
Young *Pinus radiata* (NZ) (closed canopy)	19% of 7-month summer/ autumn period	Fahey *et al.* (2001)

Note: *The figures denoted with an asterisk are actually evapotranspiration values rather than absolute interception loss, leading to higher values

series of paired catchment studies in Oregon, USA and concluded that there was clear evidence of changes in interception rates and peak discharges. Thomas and Megahan (1998) reanalysed the data used by Jones and Grant (1996) and came to the conclusion that although there was clear evidence of changes in peak flows in the small-scale catchment pairs (60–100 km²) there was no change, or inconclusive evidence for change, in the large catchments (up to 600 km²). There has followed a series of letters between the authors disputing various aspects of the studies (see *Water Resources Research* volume 37: 175–183). This debate in the research literature mirrors the overall concern in hydrology over the scale issue. There are many processes that we measure at the small hillslope level that may not be important when scaled up to larger catchments.

Land drainage

Land drainage is a common agricultural 'improvement' technique in areas of high rainfall and poor natural drainage. In an area such as the Fens of Cambridgeshire, Norfolk and Lincolnshire in England this has taken the form of drains or canals and an elaborate pumping system, so that the natural wetlands have been drained completely. The result of this has been the utilisation of the area for intensive agricultural production since the drainage took place in the seventeenth and eighteenth centuries. Since that time the land has sunk, due to the removal of water from the peat-based soils, and the area is totally dependent on the pumping network for flood protection. To maintain this network vegetation control and clearance of silt within channels is required, a cost that can be challenged in terms of the overall benefit to the community (Dunderdale and Morris, 1996).

At the smaller scale, land drainage may be undertaken by farmers to improve the drainage of soils. This is a common practice throughout temperate regions and allows soils to remain relatively dry during the winter and early spring. The most common method of achieving this is through a series of tile drains laid across a field that drain directly into a water course (often a ditch). Traditionally tile drains were clay pipes that allowed water to drain into them through the strong hydraulic gradient created by their easy drainage towards the ditch. Modern tile drains are plastic pipes with many small holes to allow water into them. Tile drains are normally laid at about 60 cm depth and should last for at least 50 years or more. To complement the tile drains 'mole drainage' is carried out. This involves dragging a large, torpedo-shaped metal 'mole' behind a tractor in lines orthogonal to the tile drains. This creates hydrological pathways, at 40–50 cm depth, towards the tile drains. Mole draining may be a regular agricultural activity, sometimes every 2–5 years in heavy agricultural land (i.e. clay soils). Normal plough depth is around 30 cm, so that the effects of mole draining last beyond a single season.

The aim of tile and mole drainage is to hold less water in a soil. This may have two effects on the overall hydrology. It allows rapid drainage from the field, therefore increasing the flashy response (i.e. rapid rise and fall of hydrograph limbs) in a river. At the same time the lack of soil moisture may lead to greater infiltration levels and hence less overland flow. Spaling (1995) notes that in Southern Ontario, Canada, land drainage alters timing and volume of water flow at the field scale, but it is difficult to detect this at the watershed scale. Hiscock *et al.* (2001) analysed sixty years of flow records for three catchments in Norfolk, UK to try and detect any change in the rainfall–runoff relationship during this time. The conclusion of their study was that despite much land drainage during the period of study the rainfall–runoff relationship 'remained essentially unchanged' (Hiscock *et al.*, 2001). This lack of change in overall hydrology, despite the known land drainage, may be due to the two hydrological impacts cancelling each other out, or else that the impact of land drainage is small, particularly at the large catchment scale.

Land drainage can be a significant factor in upland areas used for forestry. A common technique in Europe is using the plough and furrow method of drainage. A large plough creates drains in an area, with the seedlings being planted on top of the soil displaced by the plough (i.e. immediately adjacent to the drain but raised above the water table). Like all land drainage this will lower the water table and allow rapid routing of stormflow. A study at a small upland catchment in the north-east of England has shown that land drainage effects are drastic, and only after 30 years of afforestation has the impact lessened (Robinson, 1998). This long recovery time may be a reflection of the harsh environment the trees are growing in; other areas have recovered much faster.

Salination

Salination is an agricultural production problem that results from a build up of salt compounds in the surface soil. Water flowing down a river is almost never 'pure', it will contain dissolved solids in the form of salt compounds. These salt compounds are derived from natural sources such as the weathering of surface minerals and sea spray contained in rainfall. When water evaporates the salts are left behind, something we are familiar with from salt lakes such as in Utah, central Australia, and the Dead Sea in the Middle East. The same process leads to salinity in the oceans.

Salination of soils (often also referred to as salinisation) occurs when there is an excess of salt-rich water that can be evaporated from a soil. The classic situation for this is where river-fed irrigation water is used to boost agricultural production in a hot, dry climate. The evapotranspiration of salt-rich irrigation water leads to salt compounds accumulating in the soil, which in turn may lead to a loss of agricultural production as many plants fail to thrive in a salt-rich environment. Although salination is fundamentally an agronomic problem it is driven by hydrological factors (e.g. water quality, evaporation rates), hence the inclusion in a hydrological textbook.

Salination of soils and water resources have been reported from many places around the world. O'Hara (1997) provides data on waterlogging and subsequent salination in Turkmenistan, the direct result of irrigation. Gupta and Abrol (2000) describe the salinity changes that have occurred in the Indo-Gangetic Plains on the Indian sub-continent following increased rice and wheat production. Flugel (1993) provides data on irrigation return flow leading to salination of a river in the Western Cape Province of South Africa. Prichard *et al.* (1983) report salination of soils from irrigation in California, USA. Irrigation water often has a high total dissolved solids (TDS) load before being used for agricultural production. Postel (1993) suggests that typical values range from 200 to 500 mg/l, where water is considered brackish at levels greater than 300 mg/l. Postel also states that using this type of irrigation water under a normal irrigation level would add

between 2 and 5 tons of salt per hectare per year. The vast majority of this salt is washed out of the soil and continues into a water table or river system; some, though, will be retained to increase salination in the soil.

Case study

SALINATION OF WATERWAYS IN AUSTRALIA: A SALINITY PROBLEM FROM LAND USE CHANGE

Salination of surface waters is a huge problem for large areas of Australia. Sadler and Williams (1981) estimate that a third of surface water resources in south-west Western Australia can be defined as brackish (from Williamson *et al.*, 1987). In the same region it is estimated that 1.8 million hectares of agricultural land are affected by salinity problems (Nulsen and McConnell, 2000). In an assessment of ten catchments in New South Wales and Victoria (total land area of 35.7 million hectares) it is estimated that 4.1 per cent of the land area is affected by salinity and that this imposes a cost of $122 million (Australian dollars) on agricultural production (Ivey ATP, 2000).

This salinity problem is a steady increase in concentration of salt compounds in rivers, leading to the surface water becoming unusable for public supply or irrigation. In south-west Western Australia salinity levels in soils are typically between 20 and 120 kg/m^2 (Schofield, 1989). (Williamson *et al.* [1987] quote a range of 0.2–200 kg/m^2.) This is very high and is a result of thousands of years of low rainfall and high evaporation leading to an accumulation of wind-borne sea salt in the soil. The natural vegetation for this area is deep-rooted eucalypt forest and savannah woodland which has a degree of salt tolerance and the ability to extract water from deep within the soil. The ability to draw water from deep within a soil maintains a high soil water deficit which is filled by seasonal rainfall (Walker

et al., 1990). The removal of this native vegetation and replacement with shallow-rooted crops (particularly wheat) and pasture has led to a fundamental change in the hydrology, which in turn has led to a change in salinity. The replacement vegetation does not use as much water, leading to greater levels of groundwater recharge and rising water tables. This is particularly so when wheat, which is largely dormant, is planted, or when the ground is left fallow during the wetter winter season. The groundwater is often saline and in addition to this, as the water table rises, it takes up the salt stored in soils. Rivers receive more groundwater recharge (with saline water) and the streams increase in salinity. The link between vegetation change and increasing salinity levels was first proposed by Wood (1924) and has since been demonstrated through field studies.

Williamson *et al.* (1987) carried out a catchment-based study of vegetation change and increasing salinity in Western Australia. The study monitored salinity and water quantity in four small catchments, two of which were cleared of native vegetation and two kept as controls. The monitoring took place between 1974 and 1983, with the vegetation change occurring at the end of 1976 and start of 1977; a selection of results is shown here. There was a marked change in the hydrological regime (see Figure 9.3), with a large increase in the amount of streamflow as a percentage of rainfall received. This reinforces

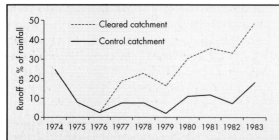

Figure 9.3 Streamflow expressed as a percentage of rainfall for two catchments in south-west Western Australia. The control maintained a natural vegetation while in the other catchment the bush was cleared during 1976/77 and replaced with pasture.
Source: Data from Williamson *et al.* (1987)

the idea of Wood (1924) that the native vegetation uses more water than the introduced pasture species.

In terms of salinity there was also a marked change, although this is not immediately evident from a time series plot (Figure 9.4). The chloride concentration in streamflow is a good indicator of salinity as it is one of the main salts that would be expected to be deposited from sea spray; however, it is not the total salinity. In Figure 9.4 there appears to be an increasing difference between

chloride concentrations with time. Chloride concentration shows considerable variation between years, which is related to variation in rainfall between years. The peaks in salinity correspond to years with high rainfall. To remove this factor Williamson *et al.* (1987) calculated the chloride concentration as a ratio between output (measured in the streamflow) and input (measured in the rainfall). This is shown in Figure 9.5.

When the chloride level is expressed as this output/input ratio (Figure 9.5) it is easy to see a marked difference following the vegetation change. In the years following 1976/77 there is considerably more output of chloride than input (i.e. the ratio is well above a value of 1), a result of the chloride being leached out of the soil. In this manner the chloride concentration in the river is staying at a high level even when there is a low input (i.e. low rainfall). Prior to vegetation change the ratio is approximately even, the chloride inputs and output had reached some type of equilibrium. Given enough time the same would happen again with the new vegetation cover, but first a large store of chloride would be released from the soil. This is a case where the vegetation change has upset the hydrological balance of a catchment, which in turn has implications for water quality.

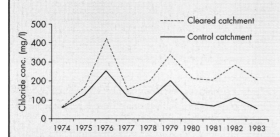

Figure 9.4 Chloride concentrations for two catchments in south-west Western Australia. These are the same two catchments as in Figure 9.3. NB World Health Organisation guidelines suggest that drinking water should have chloride concentration of less than 250 mg/l.
Source: Data from Williamson *et al.* (1987)

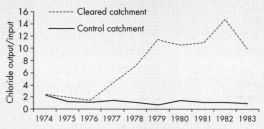

Figure 9.5 Chloride output/input ratio for two catchments in south-west Western Australia. These are the same two catchments as in Figures 9.3 and 9.4. Input has been measured through chloride concentrations in rainfall while output is in streamflow.
Source: Data from Williamson *et al.* (1987)

GROUNDWATER DEPLETION

In many parts of the world there is heavy reliance on aquifers for provision of water to a population. In England around 30 per cent of reticulated water comes from groundwater, but that rises to closer to 75 per cent in parts of south-east England. The water is extracted from a chalk aquifer that by and large receives a significant recharge during the winter months. Apart from very dry periods (e.g. the early 1990s) there is normally enough recharge to sustain withdrawals. Not all groundwater is recharged so readily. Many aquifers have built up their water reserves over millions of years and receive very little infiltrating rainfall on a year by year basis. Much of the Saudi peninsula in the Middle East is underlain by such an aquifer. The use of this water at high rates may lead to groundwater depletion, a serious long-term problem for water management. The Ogallala aquifer Case Study introduces groundwater depletion problems in the High Plains region of the USA.

Case study

OGALLALA AQUIFER DEPLETION

The Ogallala aquifer (also called the High Plains aquifer) is a huge groundwater reserve underlying an area of approximately 583,000 km^2 in the Great Plains region of the USA. It stretches from South Dakota to Texas and also underlies parts of Nebraska, Wyoming, Colorado, Kansas and New Mexico (see Figure 9.6).

The aquifer formed through erosion from the Rocky Mountains to its west. The porous material deposited from this erosion was filled with water from rivers draining the mountains and crossing the alluvial plains. This has created a water reserve that in places is 300 m deep. A major problem is that the aquifer is now isolated from the Rocky Mountains as a recharge source and has to rely on natural replenishment from local rainfall and infiltration. This is a region that receives around 380–500 mm of rainfall per annum and has very high evaporation rates during the summer. The climate is classified as semi-arid.

Ever since Europeans first settled the Great Plains the Ogallala aquifer has been an important water source for irrigation and drinking water supply. Since the 1940s there has been rapid expansion in the amount of irrigated land in the region (see Figure 9.7), so that in 1990 as much as 95 per cent of water drawn from the aquifer was used for irrigating agricultural land (McGuire and Fischer, 1999). Improving technology has meant

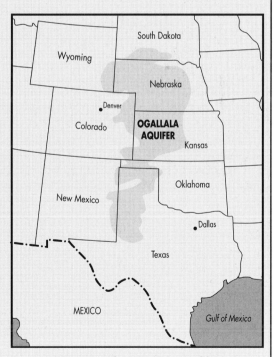

Figure 9.6 Location of the Ogallala aquifer in the Midwest of the USA.

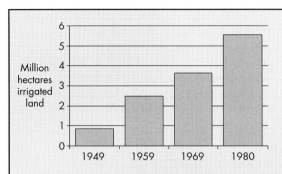

Figure 9.7 Amount of irrigated land using groundwater in the High Plains region.
Source: Data from McGuire and Fischer (1999)

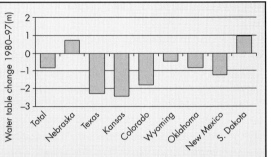

Figure 9.8 Average changes in the water table for states underlying the Ollagala aquifer. The states are arranged in size of area overlying the aquifer (Nebraska largest, South Dakota the least).
Source: Data from McGuire and Fischer (1999)

that the windmill-driven irrigation that was predominant in the 1940s and 1950s has been replaced with pumps capable of extracting vast amounts of water at a rapid rate. The result of this has been drastic declines in water tables – as much as 30 m in parts of Texas, New Mexico and Kansas (McGuire and Fischer, 1999).

There have been various efforts made to try and reduce the depletion of the Ollagala aquifer, but this is made difficult by the importance this area has for agricultural production in the USA. Systems of irrigation scheduling have been introduced to make the use of irrigated water more efficient. This involves a close monitoring of soil moisture content so that water is only applied when needed by plants and the actual amount required can be calculated. Another management tool to lessen depletion is changing agricultural production so that water-thirsty plants such as cotton are not grown in areas that rely on groundwater for irrigation.

The United States Geological Survey (USGS) has been monitoring changes in water in over 7,000 wells since the late 1980s in order to assess the rate of overall groundwater depletion. Average figures for the period 1980–97 are given in Figure 9.8. This shows an overall decline in water tables of around 0.8 m but that some regions have shown a rise (e.g. Nebraskaa 0.7 m rise). Although the overall water table has declined there has been a slowing in the rate of decline. This has been attributed to various factors including a wetter than usual period from 1980 to 1997, more efficient irrigation usage and technology, regulation of groundwater withdrawals, and changing commodity prices in agriculture.

It is encouraging that a decline in the rate of water table drops has occurred in the Ollagala aquifer, but these still represent an unsustainable depletion of the groundwater. It is difficult to see how the decline could be halted without a complete change in agricultural production for the region, but this is unlikely to occur until the price of extracting the water is too high to be economically viable. At the moment the region is using an unsustainable management practice that has led to substantial groundwater depletion; this is likely to continue into the near future.

URBAN HYDROLOGY

Many aspects of urban hydrology have already been covered, especially with respect to water quality (see Chapter 8), but the continuing rise in urban population around the world makes it an important issue to consider under the title of change. There is no question that urban expansion has a significant effect on the hydrology of any river draining the area. Initially this may be due to climate alterations affecting parts of the hydrological cycle. The most obvious hydrological impact is on the runoff hydrology, but other areas where urbanisation may have an impact are point source and diffuse pollution affecting water quality, river channelisation to control flooding, increased snow-melt from urban areas and river flow changes from sewage treatment.

Urban climate change

In Table 9.3 some of the climatic changes due to urbanisation are expressed as a ratio between the urban and rural environments. This suggests that within a city there is a 15 per cent reduction in the amount of solar radiation reaching a horizontal surface, a factor that will influence the evaporation rate. Studies have also found that the precipitation

levels in an urban environment are higher by as much as 10 per cent. Atkinson (1979) detected an increase in summer thunderstorms over London which was attributed to extra convection and condensation nuclei being available. Other factors greatly affected by urbanisation are winter fog (doubled) and winter ultraviolet radiation (reduced by 30 per cent).

Urban runoff

The changes in climate are relatively minor compared to the impact impermeable surfaces in the urban environment have on runoff hydrology. Roofs, pavement, roads, parking lots and other impermeable surfaces have extremely low infiltration characteristics, consequently Hortonian overland flow readily occurs. These surfaces are frequently linked to gutters and stormwater drains to remove the runoff rapidly. The result of this is far greater runoff and the time to peak discharge being reduced. Cherkauer (1975) compared two small catchments in Wisconsin, USA. The rural catchment had 94 per cent undeveloped land while the urban catchment had 65 per cent urban coverage. During a large storm in October 1974 (22 mm of rain in 5 hours) the peak discharge from the urban catchment area was over 250 times that of the

Table 9.3 Difference in climatic variables between urban and rural environments

Climatic variable	Ratio of city:environs
Solar radiation on horizontal surfaces	0.85
UV radiation: summer	0.95
UV radiation: winter	0.70
Annual mean relative humidity	0.94
Annual mean wind speed	0.75
Speed of extreme wind gusts	0.85
Frequency of calms	1.15
Frequency and amount of cloudiness	1.10
Frequency of fog: summer	1.30
Frequency of fog: winter	2.00
Annual precipitation	1.10
Days with less than 5 mm precipitation	1.10

Source: From Lowry (1967)

rural catchment (Cherkauer, 1975). The storm hydrograph from this event was considerably more flashy for the urban catchment (i.e. it had a shorter, sharper peak on the hydrograph).

Rose and Peters (2001) analysed a long period of streamflow data (1958–96) to detect differences between urbanised and rural catchments near Atlanta, Georgia, USA. The stormflow peaks for large storms were between 30 and 100 per cent larger in the urbanised catchment, with a considerably shorter recession limb of the hydrograph. In contrast to the stormflows, lowflows were 25–35 per cent less in the urban catchment, suggesting a lower rainfall infiltration rate. Overall there was no detectable difference in the annual runoff coefficient (runoff as percentage of precipitation) between urban and rural catchments.

Pollution from urban runoff

There is a huge amount of research and literature on the impacts of urbanisation on urban water quality. Davis *et al.* (2001) link the accumulation of heavy metals in river sediments to urban runoff, particularly from roads. Specific sources are tyre wear and vehicle brakes for zinc, and buildings for lead, copper, cadmium and zinc (Davis *et al.*, 2001). In Paris Gromaire-Mertz *et al.* (1999) found high concentrations of heavy metals in runoff from roofs, while street runoff had a high suspended solids and hydrocarbon load. The hydrocarbons are of particular concern, especially the carcinogenic polycyclic aromatic hydrocarbons (PAH) derived from petrol engines. Krein and Schorer (2000) trace PAHs from road runoff into river sediments where they bind onto fine sand and silt particles.

The nature of urban runoff (low infiltration and rapid movement of water) concentrates the pollutants in the first flush of water. Studies have shown that over 80 per cent of pollutant particles are washed into a drainage system within the first 6–10 mm of rain falling (D'Arcy *et al.*, 1998), and often from a very small collection area within the urban catchment (Lee and Bang, 2000). This information is important when proposing strategies to

deal with the urban pollutant runoff. One of the main methods is to create an artificial wetland within an urban setting so that the initial flush of storm runoff is collected, slowed down, and pollutants can be modified by biological action. Shutes (2001) has a review of artificial wetlands in Hong Kong, Malaysia and England and discusses the role of plants in improving water quality. Scholes *et al.* (1999) show that two artificial wetlands in London, England are efficient in removing heavy metals and lowering the BOD of urban runoff during storm events. Carapeto and Purchase (2000) report similar efficiency for the removal of cadmium and lead from urban runoff.

River channelisation

It is a common practice to channelise rivers as they pass through urban areas in an attempt to lessen floods in the urban environment. Frequently, although not always, this will involve straightening a river reach which has impacts on the streamflow. Simons and Senturk (1977) list some of the hydrological impacts of channel straightening: higher velocities in the channel; increased sediment transport and possible base degradation; increased stormflow stage (height); and deposition of material downstream of the straightening. The impact of urban channelisation is not restricted to the channelised zone itself. The rapid movement of water through a channelised reach will increase the velocity, and may increase the magnitude, of a flood wave travelling downstream. Deposition of sediment downstream from the channelised section may leave this area prone to flooding through a raised river bed.

Snow-melt

The influence of urbanisation on snow-melt is complicated. Semandi-Davies (1998) suggests that melt intensities are generally increased in an urban area, although shading may reduce melt in some areas. Overall there is a greater volume of water in the early thaw from an urban area when compared to rural (Taylor and Roth, 1979; Semandi-Davies, 1998). This may be complicated by snow clearance operations, particularly if the cleared snow is placed in storage areas for later melting (Jones, 1997). In this case the greater mass of snow in a small area will cause slower melt than if distributed throughout the streets.

Waste water input and water extraction

Human intervention in the hydrological regime of a river can be in the form of extraction (for irrigation or potable supply) or additional water from waste water treatment plants. The amount of water discharged from a sewage treatment works into a river may cause a significant alteration to the flow regime. At periods of low flow 44 per cent of the river Trent (a major river draining eastern England) may comprise water derived from waste water effluent (Farrimond, 1980 quoted in Newson, 1995). There are times when the river Lea (a tributary of the Thames, flowing through north-east London) is composed of completely recycled water, which may have been through more than one sewage works. Jones has a startling diagram (1997: 227, figure 7.9) showing very large diurnal variations in the river Tame that can be attributed to sewage effluent flows from the city of Birmingham, England. In this case the lowest flow occurs at 6 a.m. with a rapid rise in effluent flow that by mid-morning has boosted flow in the river Tame by around 40 per cent ($\approx$2.3 cumecs) (Jones, 1997).

The extra flow that a river derives from sewage effluent may be especially significant if the waste water effluent has been abstracted from another catchment. The water in the river Tame naturally flows into the river Trent before flowing into the North Sea on the east coast of England. A large amount of abstracted water for Birmingham comes from the Elan valley in Wales, a natural tributary of the river Wye which drains into the Bristol Channel on the west coast of England. So in addition to causing diurnal fluctuations in the river Tame downstream of Birmingham, the waste water effluent is part of a water transfer across Great Britain. In a study into low flows in the United Kingdom (i.e. England, Wales, Scotland and Northern Ireland), Gustard et al. (1992) identified that 37 per cent of flow gauges were measuring flow regimes subject to artificial influence such as abstraction, effluent discharge or reservoir regulation of the river. This degree of flow alteration is a reflection of the high degree of urbanisation and the high percentage of the population living in an urban environment in the United Kingdom.

SUMMARY

The case studies and different sections in this chapter have shown that there are many aspects of change in hydrology to be considered. In order to understand and make predictions concerning change it is essential to understand the fundamentals of hydrology: how processes operate in time and space; how to measure and estimate the rates of flux for those processes; and how to analyse the resultant data. The fundamental processes do not change, it is their rates of flux in different locations that alter. It is fundamentally important that hydrology as a science is investigating these rates of change, and finding new ways of looking at the scales of change in the next 100 years.

ESSAY QUESTIONS

1 **Explain the way that human-induced climate change may affect the hydrological regime for a region.**

2 **Assess the role of land use change as a major variable in forcing change in the hydrological regime for a region near you.**

3 **Compare and contrast the impact of urbanisation to the impact of land use change on general hydrology within the country where you live.**

4 Discuss the major issues facing water resource managers over the next 50 years in a specified geographical region.

FURTHER READING

Acreman, M. (ed.) (2000) *The hydrology of the UK, a study of change.* Routledge, London.
Chapters by different authors looking at change in the UK.

Intergovernmental Panel on Climate Change (IPCC) (2001) *Climate change 2001.* Cambridge University Press, Cambridge.
Various reports are published by the IPCC, summaries of which can be found at http://www.ipcc.ch

GLOSSARY

actual evaporation Evaporation which occurs irrespective of the amount of available water

advective energy energy that originates from elsewhere (another region that may be hundreds or thousands of kilometres away) and has been transported to a region (frequently in the form of latent heat) where it becomes available energy in the form of sensible heat.

aerodynamic resistance A term to account for the way that the atmosphere mixes with evaporating air above it through turbulent mixing of the atmosphere.

albedo The reflectivity of a surface (a unit percentage).

alkalinity A measure of the capacity to absorb hydrogen ions without a change in pH (Viessman and Hammer, 1998). This is influenced by the concentration of hydroxide, bicarbonate or carbonate ions.

amenity value A term used to denote how useful an area (e.g. a stretch of river) is for recreation and other purposes.

anemometer Instrument for measuring wind speed.

annual maximum series A form of data that may be used in flood frequency analysis.

aquifer A layer of unconsolidated or consolidated rock that is able to transmit and store enough water for extraction. A confined aquifer has restricted flow above and below it while an unconfined aquifer has no upper limit.

aquifer storage and recovery A water resource management technique involving the addition of surface water into an aquifer for storage to be recovered later.

aquifuge A totally impermeable rock formation.

aquitard A geological formation that transmits water at a much slower rate than the aquifer.

areal rainfall The average rainfall for an area (often a catchment in hydrology) calculated from several different point measurements.

artesian water or well Water that flows directly to the surface from a confined aquifer (i.e. it does not require extraction from the ground via a pump).

AVHRR (Advanced Very High Resolution Radiometer) A North American Space Agency (NASA) satellite used mainly for atmospheric interpretation.

bankfull discharge The amount of water flowing down a river when it is full to the top of its banks.

baseflow The portion of streamflow between peaks (as seen in a hydrograph) that is not attributed to storm precipitation. Sometimes also referred to as slowflow.

Bergeron process The process of raindrop growth through a strong water vapour gradient between ice crystals and small water droplets.

biochemical oxygen demand (BOD) A measure of the oxygen required by bacteria and other micro-organisms to break down organic matter in a water sample. A strong indicator of the level of organic pollution in a river.

Bowen ratio The ratio of sensible heat to latent heat. This is sometimes used within a method to measure evaporation from a surface.

Boyle's law A law of physics relating pressure (P), temperature (T), volume (V) and concentration of molecules (n) in gases. $PV = nRT$. R is a constant.

canopy storage capacity The volume of water that can be held in the canopy before water starts dripping as indirect throughfall.

capillary forces The forces holding back soil water so that it does not drain completely through a soil under gravity.

catchment The area of land from which water flows towards a river and then in that river to the sea. Also known as a river basin.

channel flow Water flowing within a channel. A general term for streamflow or riverflow.

channelisation The confinement of a river into a permanent, rigid, channel structure. This often occurs as part of urbanisation and flood protection.

cloud seeding The artificial generation of precipitation (normally rainfall) through provision of extra condensation nuclei within a cloud.

condensation The movement of water from a gaseous state into a liquid state; the opposite of evaporation.

condensation nuclei Minute particles present in the atmosphere which the water or ice droplets form upon.

convective precipitation Precipitation caused by heating from the earth's surface (leading to uplift of a moist air body).

covalent bonding A form of molecular bonding where electrons are shared between two atoms in the molecule. This is the strongest form of chemical bond and exists within a water molecule.

cyclonic precipitation Precipitation caused by a low-pressure weather system where the air is constantly being forced upwards.

dewfall (or dew) Water that condenses from the atmosphere (upon cooling) onto a surface (frequently vegetation).

dilution gauging A technique to measure streamflow based on the dilution of a tracer by the water in the stream.

discharge See **streamflow**.

effective rainfall The rainfall that produces stormflow. This is a term used in the derivation and implementation of the unit hydrograph.

eutrophication A term used to describe the addition of nutrients to an aquatic ecosystem that leads to an increase in net primary productivity. The term 'cultural eutrophication' is sometimes used to indicate the enhanced addition of nutrients through human activity. This may lead to problems with excess weed and algal growth in a river.

evaporation The movement of water from a liquid to a gaseous form (i.e. water vapour) and dispersal into the atmosphere.

evaporation pan A large pan of water, with a measuring stick or weighing device underneath that allows you to record how much water is lost through evaporation over a time period.

evapotranspiration A combination of actual evaporation and transpiration. The term recognises the fact that much of the earth's surface is a mixture of vegetation cover and bare soil.

field capacity The actual maximum water content that a soil can hold under normal field conditions. This is often less than the saturated water content as the water does not fill all the pore space and is constantly under the influence of gravity.

flash flood A flood event that occurs as a result of extremely intense rainfall causing a rapid rise in water levels in a stream. This is common in arid and semi-arid regions.

flood An inundation caused by a period of abnormally large discharge.

flood frequency analysis A technique to investigate the magnitude–frequency relationship for floods in a particular river. This is based on historical hydrograph records.

flow duration curve A graphical description of the percentage of time a certain discharge is exceeded for a particular river.

flux The rate of flow of some quantity (e.g. the rate of flow of water as evaporation is referred to as an evaporative flux).

frequency–magnitude The relationship between how often a particular size storm (or runoff) event occurs. In hydrology it is common to study low frequency–high magnitude events (e.g. large floods do not happen very often).

Geographic Information Systems (GIS) A computer program which is able to store and manipulate spatial digital data over an area (e.g. maps).

geomorphology The study of landforms.

gravimetric soil moisture content The ratio of the weight of water in a soil to the overall weight of a soil.

groundwater Water held in the saturated zone beneath a water table. This is also referred to as water in the phreatic zone.

groundwater flow Water which moves in the saturated (phreatic) zone.

hillslope hydrology The study of hydrological processes operating at the hillslope scale.

Hjulstrom curve The relationship between water velocity and sediment erosion and deposition.

Hortonian overland flow See **infiltration excess overland flow**.

hydraulic conductivity The measure of ability of a porous medium to transmit water. This is a flux term with units of metres per second. The hydraulic conductivity of a soil is highly dependent on water content.

hydraulic radius The wetted perimeter of a river divided by the cross-sectional area.

hydrogen bonding Bonding between atoms or molecules caused by the electrical attraction between a negative and positive ion. This type of bonding exists between water molecules.

hydrograph A continuous record of streamflow.

hydrograph separation The splitting of a hydrograph into stormflow and baseflow.

hydrological cycle A conceptual model of how water moves around between the earth and atmosphere in different states as a gas, liquid or solid. This can be at the global or catchment scale.

hydrology 'The science or study of' ('logy' from Latin *logia*) and 'water' ('hydro' from Greek *hudor*). Modern hydrology is concerned with the distribution of fresh water on the surface of the earth and its movement over and beneath the surface, and through the atmosphere.

hydrometry The measurement of streamflow.

hypsometric method A method for estimating areal rainfall based on the topography of your area (e.g. a catchment).

hysteresis The difference in soil suction at a given water content dependent on whether the soil is being wetted or dried.

infiltration capacity The rate of infiltration when a soil is fully saturated (i.e. at full capacity of water).

infiltration excess overland flow Overland flow that occurs when the rainfall rate exceeds the infiltration rate for a soil. Also referred to as Hortonian overland flow.

infiltration rate How much water enters a soil during a certain time interval.

infiltrometer An instrument to measure the infiltration rate and infiltration capacity for a soil.

interception The interception of precipitation above the earth's surface. This may be by a vegetation canopy or buildings. Some of this intercepted water may be evaporated; referred to as interception loss.

isohyetal method A method for estimating areal rainfall based on the known distribution of rainfall within the area (e.g. a catchment).

jökulhlaup The flood resulting from an ice-dam burst.

kriging A spatial statistics technique that identifies the similarity between adjacent and further afield point measurements. This can be used to interpolate an average surface from a series of point measurements.

LANDSAT (LAND SATellite) A series of satellites launched by the North American Space Agency (NASA) to study the earth's surface.

latent heat The energy required to produce a phase change from ice to liquid water, or liquid water to water vapour. When water moves from liquid to gas this is a negative flux (i.e. energy is lost), whereas the opposite phase change (gas to liquid) produces a positive heat flux.

lateral flow See **throughflow**.

low flow A period of extreme low flow in a river hydrograph (e.g. summer or dry season river flows).

low flow frequency analysis A technique to investigate the magnitude–frequency relationship for low flows in a particular river. This is based on historical hydrograph records.

lysimeter A cylinder filled with soil and plants used to measure evaporation from a vegetated surface. This can be done either as a weight loss or through solving some form of the water balance equation.

macropores Large pores within a soil matrix, typically with a diameter greater than 3 mm.

model A representation of the hydrological processes operating within an area (usually a catchment). This is a usually used to mean a numerical model, which simulates the flow in a river, based on mathematical representations of hydrological processes.

mole drainage An agricultural technique involving the provision of rapid subsurface drainage routes within an agricultural field.

net radiation The total electromagnetic radiation (in all wavelengths) received at a point.

neutron probe An instrument to estimate the soil water content.

open water evaporation The evaporation that occurs above a body of water such as a lake, stream or the oceans.

orographic precipitation Precipitation caused by an air mass being forced to rise over an obstruction such as a mountain range.

overland flow Water which runs across the surface of the land before reaching a stream. This is one form (but not the only form) of runoff.

oxygen sag curve The downstream dip in dissolved oxygen content that can be found after the addition of organic pollution.

partial areas concept The idea that only certain parts of a catchment area contribute overland flow to stormflow; compare to the variable source areas concept.

partial duration series A form of data that may be used in flood frequency analysis.

peakflow See **stormflow**.

perched water table Area where the water table is above a regional water table, usually due to small impermeable lenses in the geological formation.

pH The concentration of hydrogen ions within a water sample. A measure of water acidity.

phreatic zone The area beneath a water table (i.e. groundwater).

piezometer A tube with holes at the base that is placed at depth within a soil or rock mantle to measure the water pressure at a set location.

pipeflow The rapid movement of water through a hillslope in a series of linked pipes. (NB these can be naturally occurring.)

porosity The percentage of pore space (i.e. air) within a soil.

potential evaporation Evaporation which occurs over the land's surface if the water supply is unrestricted.

precipitation The movement of water from the atmosphere to the earth's surface. This can occur as rain, hail, sleet or snowfall.

quickflow See **stormflow**.

rainfall Precipitation in a liquid form. The usual expression of rainfall is as a vertical depth of water (e.g. mm or inches).

rainfall intensity The rate at which rainfall occurs. A depth of rainfall per unit time, most commonly mm per hour.

rain gauge An instrument for measuring the amount of rainfall at a point. Standard rain gauges are measured over a time interval such as a day, continuous rainfall measurement can be provided by special rain gauges such as the tipping-bucket gauge.

rain shadow effect An uneven distribution of rainfall caused by a large high landmass (e.g. a mountain range). On the downwind side of the mountain range there is often less rainfall (i.e. the mountain casts a rain shadow).

rating curve The relationship between river stage (height) and discharge.

recession limb (of hydrograph) The period after a peak of stormflow where the streamflow values gradually recede.

relative humidity How close to fully saturated the atmosphere is (a percentage – 100 per cent is fully saturated for the current temperature).

rising limb (of hydrograph) The start of a stormflow peak.

river basin The area of land from which water flows towards a river and then in that river to the sea. Also known as the catchment area.

roughness coefficient A term used in equations such as Chezy and Manning's to estimate the degree that water is slowed down by friction along the bed surface.

runoff The movement of liquid water above and below the surface of the earth.

salination The build up of salts in a soil or water body.

satellite remote sensing The interpretation of ground (or atmospheric) characteristics based on measurements of radiation from the earth/atmosphere. The radiation measurements are received on satellite-based sensors.

saturated overland flow Overland flow that occurs when a soil is completely saturated.

saturated water content The maximum amount of water that the soil can hold. It is sometimes referred to as the porosity, which assumes that the water fills all the pore space within a soil.

saturation vapour pressure The maximum vapour pressure possible (i.e. the vapour pressure exerted when a parcel of air is fully saturated). The saturation point of an air parcel is temperature-dependent and hence so is the saturation vapour pressure.

sensible heat The heat which can be sensed by instruments. This is easiest understood as the heat we feel as warmth. The sensible heat flux is the rate of flow of that sensible heat.

slowflow See **baseflow**.

snowfall Precipitation in a solid form. For hydrology it is common to express the snowfall as a vertical depth of liquid (i.e. melted) water.

snow pillow An instrument used to measure the depth of snow accumulating above a certain point.

soil heat flux Heat released from the soil having been previously stored within the soil.

soil moisture characteristic curve A measured curve describing the relationship between the capillary forces and soil moisture content. This is also called the suction moisture curve.

soil moisture deficit The amount of water required to fill the soil up to field capacity.

soil moisture tension See **soil suction**.

soil suction A measure of the strength of the capillary forces. This is also called the moisture tension or soil water tension.

soil water Water in the unsaturated zone occurring above a water table. This is also referred to as water in the vadose zone.

specific heat capacity The amount of energy required to raise the temperature of a substance by a single degree.

SPOT French satellite to study the earth's surface.

stage The water level height of a river.

stemflow Rainfall that is intercepted by stems and branches, and flows down the tree trunk into the soil.

stomatal or canopy resistance The restriction a plant places on its transpiration rate through opening and closing stomata in the leaves.

storage A term in the water balance equation to account for water that is not a flux or is very slow moving. This may include snow and ice, groundwater and lakes.

storm duration The length of time between rainfall starting and ending within a storm.

stormflow The portion of streamflow (normally seen in a hydrograph) that can be attributed to storm precipitation. Sometimes also referred to as quickflow or peakflow.

streamflow Water flowing within a stream channel (or river flow for a larger body of water). Often referred to as discharge.

suction moisture curve See **soil moisture characteristic curve**.

Synthetic Aperture Radar (SAR) A remote sensing technique that uses radar properties, usually of microwaves.

synthetic unit hydrograph A unit hydrograph derived from knowledge of catchment characteristics rather than historical hydrograph records.

tensiometer An instrument used to measure the soil moisture tension.

Thiessen's polygons A method of estimating average rainfall for an area based on the spatial distribution of rain gauges.

throughfall The precipitation that falls to the ground either directly (through gaps in the canopy), or indirectly (having dripped off leaves, stems or branches).

throughflow Water which runs to a stream in the unsaturated (vadose) zone. This is one form of runoff. Sometimes referred to as lateral flow.

time domain reflectometry (TDR) A method to estimate the soil water content.

total dissolved solids (TDS) The amount of solids dissolved within a water sample. This is closely related to the electrical conductivity of a water sample.

total suspended solids (TSS) The amount of solids suspended within a water sample. This is closely related to the turbidity of a water sample.

transpiration The movement of liquid water in a plant leaf to water vapour in the atmosphere. Plants carry out transpiration as part of the photosynthetic process.

turbidity The cloudiness of a water sample.

ultrasonic flow gauge An instrument that measures stream discharge based on the alteration to a propagated wave.

unit hydrograph A model of stormflow in a particular catchment used to predict possible future storm impacts. It is derived from historical hydrograph records.

vadose zone Area between the water table and the earth surface. The soil/rock is normally partially saturated.

vapour pressure Pressure exerted within the parcel of air by having the water vapour present within it. The more water vapour present the greater the vapour·pressure.

vapour pressure deficit The difference between the actual vapour pressure and the saturation vapour pressure.

variable source areas concept The idea that only certain parts of a catchment area contribute overland flow to stormflow and that these vary in space and time; compare to the partial areas concept.

velocity–area method A technique to measure instantaneous streamflow through measuring the cross-sectional area and the velocity through the cross section.

volumetric soil moisture content The ratio of the volume of water in a soil to the overall volume of a soil.

water balance equation A mathematical description of the hydrological processes operating within a given timeframe. Normally includes precipitation, runoff, evaporation and change in storage.

water table The surface that differentiates between fully saturated and partially saturated soil/rock.

water vapour Water in a gaseous form.

well A tube with permeable sides all the way up so that water can enter or exit from anywhere up the column. Wells are commonly used for water extraction and monitoring the water table in unconfined aquifers.

wetted perimeter The total perimeter of a cross section across a river.

wilting point The soil water content when plants start to die back (wilt).

zero plane displacement The height within a canopy at which wind speed drops to zero.

REFERENCES

Abbott, M.B., Bathurst, J.C., Cunge, J.A., O'Connell, P.E. and Rasmussen, J. (1986) An introduction to the European Hydrological System – Système Hydrologique Européen, 'SHE', 1. History and philosophy of a physically-based distributed modelling system. *Journal of Hydrology* 87:45–59.

Acreman, M. (ed.) (2000) *The hydrology of the UK, a study of change.* Routledge, London.

Alabaster, J.S. and Lloyd, R. (eds) (1980) *Water quality criteria for freshwater fish.* Butterworth and Co. Ltd, London.

Alloway, B.J. and Ayres, D.C. (1997) *Chemical principles of environmental pollution* (2nd edition). Blackie, London.

Anderson, M.G. and Burt T.P. (1978) The role of topography in controlling throughflow generation. *Earth Surface Processes and Landforms* 3:27–36.

Arnell, N.W. and Reynard, N.S. (1996) The effects of climate change due to global warming on river flows in Great Britain. *Journal of Hydrology* 183: 397–424.

Arnell, N.W. and Reynard, N.S. (2000) Climate change and UK hydrology. In: M. Acreman (ed.) *The hydrology of the UK, a study of change.* Routledge, London, pp. 3–29.

Arora, V.K. and Boer, G.J. (2001) Effects of simulated climate change on the hydrology of major river basins. *Journal of Geophysical Research – Atmospheres* 106:3335–3348.

Atkinson, B.W. (1979) Urban influences on precipitation in London. In: G.E. Hollis (ed.) *Man's influence on the hydrological cycle in the United Kingdom.* Geobooks, Norwich, pp. 123–133.

Bailey, T.C. and Gatrell, A.C. (1995) *Interactive spatial data analysis.* Longman, Harlow.

Beltaos, S. (2000) Advances in river ice hydrology. *Hydrological Processes* 14:1613–1625.

Ben-Zvi, A. (1988) Enhancement of runoff from a small watershed by cloud seeding. *Journal of Hydrology* 101:291–303.

Betson, R.P. (1964) What is watershed runoff? *Journal of Geophysical Research* 69:1541–1552.

Beven, K. (1989) Changing ideas in hydrology – the case of physically-based models. *Journal of Hydrology* 105:157–172.

Bosch, J.M. and Hewlett, J.D. (1982) A review of catchment experiments to determine the effect of vegetation changes on water yield and evapo-transpiration. *Journal of Hydrology* 55:3–23.

Brammer, D.D. and McDonnell, J.J. (1996) An evolving perceptual model of hillslope flow at the Maimai catchment. In: M.G. Anderson and S.M. Brooks (eds) *Advances in hillslope processes*. J. Wiley & Sons, Chichester, pp. 35–60.

Brandt, M., Bergstrom, S. and Gardelin, M. (1988) Modelling the effects of clearcutting on runoff – examples from central Sweden. *Ambio* 17:307–313.

Bruce, J.P. and Clark, R.H. (1980) *Introduction to hydrometeorology*. Pergamon, Toronto.

Burke, E.J., Banks, A.C. and Gurney, R.J. (1997) Remote sensing of soil–vegetation–atmosphere transfer processes. *Progress in Physical Geography* 21:549–572.

Burt, T.P. (1987) Measuring infiltration capacity. *Geography Review* 1:37–39.

Calder, I.R. (1990) *Evaporation in the uplands*. J. Wiley & Sons, Chichester.

Calder, I.R. and Newson, M.D. (1979) Land use and upland water resources in Britain – a strategic look. *Water Resources Bulletin* 15:1628–1639.

Campbell, D.I. and Murray, D.L. (1990) Water balance of snow tussock grassland in South Island, New Zealand. *Journal of Hydrology* 118:229–245.

Carapeto, C. and Purchase, D. (2000) Distribution and removal of cadmium and lead in a constructed wetland receiving urban runoff. *Bulletin of Environmental Contamination and Toxicology* 65: 322–329.

Changnon, S.A., Gabriel, K.R., Westcott, N.E. and Czys, R.R. (1995) Exploratory analysis of seeding effects on rainfall – Illinois 1989. *Journal of Applied Meteorology* 34:1215–1224.

Chapman, D. (ed.) (1996) *Water quality assessments: a guide to the use of biota, sediments and water in environmental monitoring* (2nd edition). Chapman and Hall, London.

Cherkauer, D.S. (1975) Urbanization's impact on water quality during a flood in small watersheds. *Water Resources Bulletin* 11:987–998.

Chiew, F.H.S., Whetton, P.H., McMahon, T.A. and Pittock, A.B. (1995) Simulation of the impacts of climate change on runoff and soil moisture in Australian catchments. *Journal of Hydrology* 167: 121–147.

Chow, V.T., Maidment, D.R. and Mays, L.W. (1988) *Applied hydrology*. McGraw-Hill, New York.

Christie, F. and Hanlon, J. (2001) *Mozambique & the great flood of 2000*. The International African Institute in association with James Currey (Oxford) and Indiana University Press (Bloomington and Indianapolis).

Clarke, R.T., Leese, M.N. and Newson, A.J. (1973) *Analysis of Plynlimon raingauge networks: April 1971–March 1973*. Institute of Hydrology report number 27.

Clothier, B.E., Kerr, J.P., Talbot, J.S. and Scotter, D.R. (1982) Measured and estimated evapotranspiration from well water crops. *New Zealand Journal of Agricultural Research* 25:302–307.

Cluckie, I.D. and Collier, C.G. (eds) (1991) *Hydrological applications of weather radar*. Ellis Harwood, Chichester.

Crockford, R.H. and Richardson, D.P. (1990) Partitioning of rainfall in a eucalypt forest and pine plantation in southeastern Australia. IV. The relationship of interception and canopy storage capacity, the interception of these forests and the effect on interception of thinning the pine plantation. *Hydrological Processes* 4:164–188.

D'Arcy, B.J., Usman, F., Griffiths, D. and Chatfield, P. (1998) Initiatives to tackle diffuse pollution in the UK. *Water Science and Technology* 38:131–138.

Darcy, H. (1856) *Les fontaines publiques de la ville de Dijon*. V. Dalmont, Paris.

Davie, T.J.A. (1996) Modelling the influence of afforestation on hillslope storm runoff. In: M.G. Anderson and S.M. Brooks (eds) *Advances in hillslope processes*, Vol. 1. J. Wiley & Sons, Chichester, pp. 149–184.

Davie, T.J.A. and Durocher, M.G. (1997) A model to consider the spatial variability of rainfall partitioning within deciduous canopy. I. Model description. *Hydrological Processes* 11:1509–1523.

Davie, T., Kelly, R. and Timoncini, M. (2001) SAR imagery used for soil moisture monitoring: the potential. *Remote Sensing and Hydrology 2000* (Proceedings of a symposium held at Santa Fe, New Mexico, USA, April 2000). IAHS publication number 267:327–332.

Davis, A.P., Shokouhian, M. and Ni, S.B. (2001) Loading estimates of lead, copper, cadmium, and zinc in urban runoff from specific sources. *Chemosphere* 44:997–1009.

Doorenbos, J. and Pruitt, K.C. (1975) *Crop water requirements.* FAO Irrigation and Drainage Paper 24. Rome.

Dunderdale, J.A.L. and Morris, J. (1996) The economics of aquatic vegetation removal in rivers and land drainage systems. *Hydrobiologia* 340: 157–161.

Dunne, T. (1978) Field studies of hillslope flow processes. In: M.J. Kirkby (ed.) *Hillslope hydrology.* J. Wiley & Sons, Chichester.

Dunne, T. and Black, R.D. (1970) Partial area contributions to storm runoff in a small New England watershed. *Water Resources Research* 6: 1296–1311.

Durocher, M.G. (1990) Monitoring spatial variability of rainfall interception by forest. *Hydrological Processes* 4:215–229.

Eureau (2001) Keeping raw drinking water resources safe from pesticides. Unpublished Eureau position paper EU1-01-56.

Fahey, B. and Jackson, R. (1997) Hydrological impacts of converting native forests and grasslands to pine plantations, South Island, New Zealand. *Agricultural and Forest Meteorology* 84:69–82.

Fahey, B., Watson A. and Payne, J. (2001) Water loss from plantations of Douglas-fir and radiata pine on the Canterbury Plains, South Island, New Zealand. *Journal of Hydrology* (NZ) 40(1):77–96.

Ferguson, R.I. (1999) Snowmelt runoff models. *Progress in Physical Geography* 23:205–228.

Fitzharris, B.B. and McAlevey, B.P. (1999) Remote sensing of seasonal snow cover in the mountains of New Zealand using satellite imagery. *Geocarto International* 14:33–42.

Flugel, W.A. (1993) River salination due to non-point contribution of irrigation return-flow in the Breede river, Western Cape Province, South Africa. *Water Science and Technology* 28:193–197.

Freeze, R.A. and Cherry, J.A. (1979) *Groundwater.* Prentice-Hall, Englewood Cliffs, N.J.

Freeze, R.A. and Harlan, R.L. (1969) Blueprint for a physically-based digitally-simulated hydrological response model. *Journal of Hydrology* 9:43–55.

Fritz, P., Cherry, J.A., Weyer, K.U. and Sklash, M. (1976) Runoff analyses using environmental isotopes and major ions. In: *Interpretation of environmental isotopes and hydrological data in ground water hydrology.* International Atomic Energy Agency, Vienna, pp. 111–130.

Gagin, A. and Neumann, J. (1981) The second Israeli randomized cloud seeding experiment: evaluation of results. *Journal of Applied Meteorology* 20:1301–1311.

Gardner, W.H. (1986) Water content. In: A. Klute (ed.) *Methods of soil analysis. Part 1. Physical and mineralogical methods.* American Society of Agronomy–Soil Science Society of America, Madison, Wisc., pp. 493–544.

Gleick, P.H. (1993) *Water in crisis: a guide to the world's fresh water resources.* Oxford University Press, New York.

Gordon, N.D., McMahon, T.A. and Finlayson, B.L. (1992) *Stream hydrology: an introduction for ecologists.* J. Wiley & Sons, New York.

Goudie, A., Atkinson, B.W., Gregory, K.J., Simmons, I.G., Stoddart, D.R. and Sugden, D. (eds)

(1994) *The encyclopedic dictionary of physical geography* (2nd edition). Blackwell Reference, Oxford.

Gray, N.F. (1999) *Water technology: an introduction for environmental scientists and engineers*. Arnold, London.

Grayson, R.B., Moore, I.D. and McMahon, T.A. (1992) Physically based hydrologic modelling. 2. Is the concept realistic? *Water Resources Research* 26:2659–2666.

Griffiths, G.H. and Wooding, M.G. (1996) Temporal monitoring of soil moisture using ERS-1 SAR data. *Hydrological Processes* 10:1127–1138.

Gromaire-Mertz, M.C., Garnaud, S., Gonzalez, A. and Chebbo, G. (1999) Characterisation of urban runoff pollution in Paris. *Water Science and Technology* 39:1–8.

Gupta, R.K. and Abrol, I.P. (2000) Salinity build up and changes in the rice–wheat system of the Indo-Gangetic Plains. *Experimental Agriculture* 36:273–284.

Gustard, A., Bullock, A. and Dixon, J.M. (1992) *Low flow estimation in the United Kingdom*. Publication number 108. Institute of Hydrology, Wallingford, Oxfordshire.

Harr, R.D. (1982) Fog drip in the Bull Run municipal watershed, Oregon. *Water Resources Bulletin* 18:785–789.

Hamilton, R.S. and Harrison, R.M. (1991) *Highway pollution*. Studies in Environmental Science 44. Elsevier, Amsterdam.

Hayward, D. and Clarke, R.T. (1996) Relationship between rainfall, altitude and distance from the sea in the Freetown Peninsula, Sierra Leone. *Hydrological Sciences Journal* 41:377–384.

Hedstrom, N.R. and Pomeroy, J.W. (1998) Measurements and modelling of snow interception in the boreal forest. *Hydrological Processes* 12: 1611–1625.

Heppell, C.M., Burt T.P., Williams R.J. and Haria A.H. (1999) The influence of hydrological pathways on the transport of the herbicide, isoproturon, through an under drained clay soil. *Water Science and Technology* 39:77–84.

Hewlett, J.D. and Hibbert, A.R. (1967) Factors affecting the response of small watersheds to precipitation in humid areas. In: W.E. Sopper and H.W. Lull (eds) *Forest hydrology*. Pergamon, New York, pp. 275–290.

Hewlett, J.D. and Nutter, W.L. (1969) *An outline of forest hydrology*. University of Georgia Press, Athens, Ga.

Hiscock, K.M., Lister, D.H., Boar, R.R. and Green, F.M.L. (2001) An integrated assessment of long-term changes in the hydrology of three lowland rivers in eastern England. *Journal of Environmental Management* 61:195–214.

Horton, J.H. and Hawkins, R.H. (1965) Flow path of rain from the soil surface to the water table. *Soil Science* 100:377–383.

Horton, R.E. (1933) The role of infiltration in the hydrological cycle. *Transactions of the American Geophysical Union* 14:446–460.

Howard, A. (1994) Problem cyanobacterial blooms: explanation and simulation modelling. *Transactions of the Institute of British Geographers* 19:213–224.

Hursh, C.R. (1944) Report on the sub-committee on subsurface flow. *Transactions of the American Geophysical Union* 25:743–746.

Ingwersen, J.B. (1985) Fog drip, water yield and timber harvesting in the Bull Run municipal watershed, Oregon. *Water Resources Bulletin* 21:469–473.

IPCC (Intergovernmental Panel on Climate Change) (2001) *Climate change 2001* (3 reports). Cambridge, Cambridge University Press.

Ivey ATP (2000) The cost of dryland salinity to agricultural landholders in selected Victorian and New South Wales catchments. Interim report – Part 2.

James, A. (1993) Simulation. In: A. James (ed.) *An introduction to water quality modelling* (2nd edition). J. Wiley & Sons, Chichester.

Jobling, S. and Sumpter, J.P. (1993) Detergent components in sewage effluent are weakly oestrogenic to fish: an *in vitro* study using rainbow trout (*Oncorhynchus mykiss*) hepatocytes. *Aquatic Toxicology* 27:361–372.

Jones, J.A. (2000) Hydrologic processes and peak discharge response to forest removal, regrowth, and roads in 10 small experimental basins, western Cascades, Oregon. *Water Resources Research* 36: 2621–2642.

Jones, J.A. and Grant, G.E. (1996) Peak flow responses to clear-cutting and roads in small and large basins, western Cascades, Oregon. *Water Resources Research* 32:959–974.

Jones, J.A.A. (1981) *The nature of soil piping – a review of research*. Geobooks, Norwich.

Jones, J.A.A. (1997) *Global hydrology: processes, resources and environmental management*. Longman, Harlow.

Kaleris, V., Papanastasopoulos, D. and Lagas, G. (2001) Case study of atmospheric circulation changes on river basin hydrology: uncertainty aspects. *Journal of Hydrology* 245:137–152.

Kidd, C., Kniveton, D., Barrett, E.C. (1998) The advantages and disadvantages of statistically derived, empirically calibrated passive microwave algorithms for rainfall estimation. *Journal of Atmospheric Sciences* 55:1576–1582.

Kirby, C., Newson, M.D. and Gilman, K. (1991) *Plynlimon research: the first two decades*. Institute of Hydrology report number 109. Wallingford, Oxfordshire.

Klute, A. (1986) Water retention: laboratory methods. In: A. Klute (ed.) *Methods of soil analysis. Part 1. Physical and mineralogical methods*. American Society of Agronomy–Soil Science Society of America, Madison, Wisc., pp. 635–662.

Klute, A. and Dirksen, C. (1986) Hydraulic conductivity and diffusivity: laboratory methods. In: A. Klute (ed.) *Methods of soil analysis. Part 1. Physical and mineralogical methods*. American Society

of Agronomy–Soil Science Society of America, Madison, Wisc., pp. 687–734.

Krein, A. and Schorer, M. (2000) Road runoff pollution by polycyclic aromatic hydrocarbons and its contribution to river sediments. *Water Research* 34:4110–4115.

Law, F. (1956) The effect of afforestation upon the yield of water catchment areas. *Journal of British Waterworks Association* 38:489–494.

Lee, J.H. and Bang, K.W. (2000) Characterization of urban stormwater runoff. *Water Research* 34: 1773–1780.

Lee, R. (1980) *Forest hydrology*. Columbia University Press, New York.

Leyton, L. and Carlisle, A. (1959) Measurement and interpretation of interception by forest stands. *International Association of Hydrological Sciences* 48: 111–119.

Lloyd, C.R., Gash, J.H.C., Shuttleworth, W.J. and Marques, D de O. (1988) The measurement and modelling of rainfall interception by Amazonian rainforest. *Agricultural Forestry Meteorology* 42:63–73.

Lowry, W.P. (1967) The climate of cities. *Scientific American* 217(2):20.

Lundberg, A. and Halldin, S. (2001) Snow interception evaporation. Review of measurement techniques, processes and models. *Theoretical and Applied Climatology* 70:117–133.

McDonald, A.T. and Kay, D. (1988) *Water resources: issues and strategies*. Longman, Harlow.

McDonnell, J.J. (1990) A rationale for old water discharge through macropores in a steep, humid catchment. *Water Resources Research* 26:2821–2832.

McGuire, V.L. and Fischer, B.C. (1999) *Water level changes, 1980 to 1997, and saturated thickness, 1996–97, in the High Plains aquifer*. USGS (United States Geological Survey) Fact Sheet 124–99.

McIlroy, I.C. and Angus, D.E. (1964) Grass, water and soil evaporation at Aspendale. *Agricultural Meteorology* 1:201–224.

McKim, H.L., Cassell, E.A. and LaPotin, P.J. (1993) Water resource modelling using remote sensing and object oriented simulation. *Hydrological Processes* 7:153–165.

McMillan, W.D. and Burgy, R.D. (1960) Interception loss from grass. *Journal of Geophysical Research* 65:2389–2394.

Mark, A.F., Rowley, J. and Holdsworth, D.K. (1980) Water yield from high altitude snow tussock grassland in Central Otago. *Tussock Grasslands and Mountain Lands Review* 38:21–33.

Martinec, J., Siegenthaler, H., Oescheger, H. and Tongiorgi, E. (1974) New insight into the runoff mechanism by environmental isotopes. *Proceedings of Symposium on Isotope Techniques in Groundwater Hydrology*. International Atomic Energy Agency, Vienna, pp. 129–143.

Massman, W.J. (1983) The derivation and validation of a new model for the interception of rainfall by forests. *Agricultural Meteorology* 28: 261–286.

Mather, G.K. (1991) Coalescence enhancement in large multicell storms caused by the emissions from a Kraft paper mill. *Journal of Applied Meteorology* 30:1134–1146.

Mather, G.K., Terblanche, D.E., Steffens, F.E. and Fletcher, L. (1997) Results of the South African cloud-seeding experiments using hygroscopic flares. *Journal of Applied Meteorology* 36:1433–1447.

Mauser, W. and Schädlich, S. (1998) Modelling the spatial distribution of evapotranspiration on different scales using remote sensing data. *Journal of Hydrology* 212–213:250–267.

Meybeck, M. (1981) Pathways of major elements from land to ocean through rivers. In: J.M. Martin, J.B. Burton and D. Eisma (eds) *River inputs to ocean systems*. United Nations Press, New York.

Meybeck, M., Chapman, D.V. and Helmer, R. (eds) (1989) *Global freshwater quality: a first assessment*. Blackwell, Oxford.

Middelkoop, H., Daamen, K., Gellens, D., Grabs, W., Kwadijk, J.C.J., Lang, H., Parmet, B.W.A.H., Schadler, B., Schulla, J. and Wilke, K. (2001) Impact of climate change on hydrological regimes and water resources management in the Rhine basin. *Climatic Change* 49:105–128.

Milliman, J.D. and Meade, R.H. (1983) Worldwide delivery of river sediment to the oceans. *Journal of Geology* 91:1–21.

Montagnani, D.B., Puddefoot, J., Davie, T.J.A. and Vinson, G.P. (1996) Environmentally persistent oestrogen-like substances in UK river systems. *Journal of the Chartered Institute of Water and Environmental Management* 10:399–406.

Monteith, J.L. (1965) Evaporation and environment. *Proceedings of Symposium on Experimental Biology* 19:205–234.

Mosley, M.P. (1979) Streamflow generation in a forested watershed, New Zealand. *Water Resources Research* 15:795–806.

Mosley, M.P. (1982) Subsurface flow velocities through selected forest soils, Suth Island, New Zealand. *Journal of Hydrology* 55:65–92.

Neal, C., Robson, A.J., Hall, R.L., Ryland, G., Conway, T. and Neal, M. (1991) Hydrological impacts of lowland plantations in lowland Britain: preliminary findings of interception at a forest edge, Black Wood, Hampshire, Southern England. *Journal of Hydrology* 127:349–365.

NERC (Natural Environment Research Council) (1975) *Flood studies report*. HMSO, London.

Nešpor, V. and Sevruk, B. (1999) Estimation of wind-induced error of rainfall gauge measurements using a numerical simulation. *Journal of Atmospheric and Oceanic Technology* 16:450–464.

Newson, M. (1995) Patterns of freshwater pollution. In: M. Newson (ed.) *Managing the human impact on the natural environment: patterns and processes*. J.Wiley & Sons, Chichester, pp. 130–149.

Nulsen, R. and McConnell, C. (2000) *Salinity at a glance*. Farm note can be found at: www.agric.wa.gov.au/agency/Pubns/farmnote/2000/f00800.htm

O'Hara, S.L. (1997) Irrigation and land degradation: implications for agriculture in Turkmenistan, central Asia. *Journal of Arid Environments* 37: 165–179.

Oke, T.R. (1987) *Boundary layer climates* (2nd edition). Methuen, London.

Parry, M. (1990) *Climate change and world agriculture*. Earthscan, London.

Pearce, A.J., Rowe, L.K. and O'Loughlin, C.L. (1984) Hydrology of mid-altitude tussock grass-lands, upper Waipori catchment, Otago. II. Water balance, flow duration and storm runoff. *Journal of Hydrology (NZ)* 23:60–72.

Pearson, C.P. (1998) Changes to New Zealand's national hydrometric network in the 1990s. *Journal of Hydrology (NZ)* 37:1–17.

Penman, H.L. (1948) Natural evaporation from open water, bare soil and grass. *Proceedings of the Royal Society, Series A* 193:120–145.

Penman, H.L. (1963) *Vegetation and hydrology*. Technical Communication 53, Commonwealth Bureau of Soils, Harpenden, England.

Penman, H.L. and Scholfield, R.K. (1951) Some physical aspects of assimilation and transpiration. *Proceedings of Symposium Society on Experimental Biology* 5:12–25.

Philip, J.R. (1957) The theory of infiltration. 4. Sorptivity and algebraic infiltration equations. *Soil Science* 84:257–264.

Postel, S. (1993) Water and agriculture. In: P.H. Gleick (ed.) *Water in crisis: a guide to the world's fresh water resources*. Oxford University Press, New York, pp. 56–66.

Price, M. (1996) *Introducing groundwater* (2nd edition). Chapman and Hall, London.

Priestly, C.H.B. and Taylor, R.J. (1972) On the assessment of surface heat flux using large scale parameters. *Monthly Weather Review* 100:81–92.

Prichard, T.L., Hoffman, G.J. and Meyer, J.L. (1983) Salination of organic soils in the Sacremento–San-Joaquin delta of California. *Irrigation Science* 4:71–80.

Ragan, R.M. (1968) An experimental investigation of partial area contribution. *International Association of Hydrological Sciences Publication* 76:241–249.

Rangno, A.L. and Hobbs, P.V. (1995) A new look at the Israeli cloud seeding experiments. *Journal of Applied Meteorology* 34:1169–1193.

Ranzi, R., Grossi, G. and Bacchi, B. (1999) Ten years of monitoring areal snowpack in the Southern Alps using NOAA-AVHRR imagery, ground measurements and hydrological data. *Hydrological Processes* 13:2079–2095.

Rasmussen, K.R. and Rasmussen, S. (1984) The summer water balance in a Danish Oak stand. *Nordic Hydrology* 15:213–222.

Rawls, W.J., Brakensiek, D.L. and Saxton, K.E. (1982) Estimation of soil water properties. *Transactions of the American Society of Agricultural Engineers* 25:1316–1320.

Richards, K. (1982) *Rivers form and process in alluvial channels*. Methuen, London.

Robinson, M. (1998) 30 years of forest hydrology changes at Coalburn: water balance and extreme flows. *Hydrology and Earth System Sciences* 2:233–238.

Rodda, H.J.E., Wilcock, R.J., Shankar, U. and Thorpold, B.S. (1999) The effects of intensive dairy farming on stream water quality in New Zealand. In: L. Heathwaite (ed.) *Impact of land-use change on nutrient loads from diffuse sources*. IAHS publication number 257. Wallingford, Oxfordshire.

Rodda, J.C. and Smith, S.W. (1986) The significance of the systematic error in rainfall measurement for assessing atmospheric deposition. *Atmospheric Environment* 20:1059–1064.

Rose, S. and Peters, N.E. (2001) Effects of urbanization on streamflow in the Atlanta area (Georgia, USA): a comparative hydrological approach. *Hydrological Processes* 15:1441–1457.

Ross, J.S. and Wilson, K.J.W. (1981) *Foundations of anatomy and physiology* (5th edition). Churchill Livingstone, London.

Rowntree, P.R. (1991) Atmospheric parameterisation schemes for evaporation over land: basic concepts and climate modelling aspects. In T.J. Schmugge and J.-C. André (eds) *Land surface evaporation – measurement and parameterisation*. Springer-Verlag, New York, pp. 5–29.

Russel, M.A. and Maltby, E. (1995) The role of hydrologic regime on phosphorous dynamics in a seasonally waterlogged soil. In: J.M.R. Hughes and A.L. Heathwaite (eds) *Hydrology and hydrochemistry of British wetlands*. J. Wiley & Sons, Chichester.

Rutter, A.J. (1967) An analysis of evaporation from a stand of Scots pine. In: W.E. Sopper and H.W. Lull (eds) *Forest hydrology*. Pergamon, New York, pp. 403–417.

Rutter, A.J., Kershaw, K.A., Robins, P.C. and Morton, A.J. (1971) A predictive model of rainfall interception in forests I. Derivation of the model from observations in a plantation of Corsican Pine. *Agricultural Meteorology* 9:367–384.

Rutter, A.J., Morton, A.J. and Robins, P.C. (1975) A predictive model of rainfall interception in forests II. Generalization of the model and comparison with observations in some coniferous and hardwood stands. *Journal of Applied Ecology* 12:367–380.

Sadler, B.S. and Williams, P.J. (1981) The evolution of a regional approach to salinity management in Western Australia. *Agricultural Water Management* 4:353–382.

Schofield, N.J. (1989) Stream salination and its amelioration in south-west Western Australia. *Proceedings of the Baltimore Symposium, May 1989*. IAHS publication number 182:221–230.

Scholes, L.N.L., Shutes, R.B.E., Revitt, D.M., Purchase, D. and Forshaw, M. (1999) The removal of urban pollutants by constructed wetlands during wet weather. *Water Science and Technology* 40:333–340.

Semandi-Davies, A. (1998) Modelling snowmelt induced waste water inflows. *Nordic Hydrology* 29:285–302.

Shaw, E.M. (1994) *Hydrology in practice* (3rd edition). Chapman and Hall, London.

Sherman, L.K. (1932) Streamflow from rainfall by the unit-graph method. *Engineering News Record* 108:501–505.

Shiklomanov, I.A. (1993) World fresh water resources. In: P.H. Gleick (ed.) *Water in crisis: a guide to the world's fresh water resources*. Oxford University Press, New York, pp. 13–24.

Shiklomanov, I.A. and Sokolov, A.A. (1983) Methodological basis of world water balance investigation and computation. In: A. Van der Beken and A. Herrman (eds) *New approaches in water balance computations*. IAHS publication number 148. Wallingford, Oxfordshire.

Shutes, R.B.E. (2001) Artificial wetlands and water quality improvement. *Environment International* 26:441–447.

Shuttleworth, J.W. (1988) Macrohydrology: the new challenge for process hydrology. *Journal of Hydrology* 100:31–56.

Simons, D.B. and Senturk, F. (1977) *Sediment transport technology*. Water resources publications, Fort Collins, Colo.

Sinclair, M.R., Wratt, D.S., Henderson, R. and Gray, W.R. (1996) Factors affecting the distribution and spillover of precipitation in the Southern Alps of New Zealand – a case study. *Journal of Applied Meteorology* 36:428–442.

Sklash, M.G. and Farvolden, R.N. (1979) The role of groundwater in storm runoff. *Journal of Hydrology* 43:45–65.

Slatyer, R.O. and McIlroy, I.C. (1961) *Practical micrometeorology*. CSIRO, Melbourne.

Smith, K. (2001) *Environmental hazards: assessing risk and reducing disaster* (3rd edition). Routledge, London.

Smithers, J.C., Schulze, R.E., Pike, A. and Jewitt, G.P.W. (2001) A hydrological perspective of the February 2000 floods: a case study in the Sabie river catchment. *Water SA* 27:325–332.

Spaling, H. (1995) Analyzing cumulative environmental effects of agricultural land drainage in southern Ontario, Canada. *Agriculture Ecosystems & Environment* 53:279–292.

Stewart, J.B. (1977) Evaporation from the wet canopy of a pine forest. *Water Resources Research* 13:915–921.

Stumm, W. (1986) Water and integrated ecosystem. *Ambio* 15:201–207.

Szeicz, G., Enrodi, G. and Tajchman, S. (1969) Aerodynamic and surface factors in evaporation. *Water Resources Research* 5:380–394.

Tanaka, T. (1992) Storm runoff processes in a small forested drainage basin. *Environmental Geology & Water Science* 19:179–191.

Taylor, C.H. and Roth, D.M. (1979) Effects of suburban construction on the contributing zones in a small southern Ontario drainage basin. *Hydrological Sciences Bulletin* 24:289–301.

Tebbutt, T.H.Y. (1993) *Principles of water quality control* (4th edition). Pergamon, Oxford.

Thomas, R.B. and Megahan, W.F. (1998) Peak flow responses to clear-cutting and roads in small and large basins, western Cascades, Oregon: a second opinion. *Water Resources Research* 34:3393–3403.

Thornthwaite, C.W. (1944) A contribution to the report of the committee on transpiration and evaporation, 1943–44. *Transactions of the American Geophysical Union* 25:686–693.

Thornthwaite, C.W. (1954) A re-examination of the concept and measurement of potential evapotranspiration. *Johns Hopkins University Publications in Climatology* 7:200–209.

Todd, M.C. and Bailey, J.O. (1995) Estimates of rainfall over the United Kingdom and surrounding seas from the SSM/I using the polarization corrected temperature algorithm. *Journal of Applied Meteorology* 34:1254–1265.

USCE (United States Corps of Engineers) (1996) The great flood of 1993 post-flood report. Can be found at: http://www.mvr.usace.army.mil/ PublicAffairsOffice/HistoricArchives/Floodof1993/ pafr.htm

Van Bavel, C.H.M. (1966) Potential evaporation: the combination concept and its experimental verification. *Water Resources Research* 2:455–467.

Viessman, W. Jnr. and Hammer, M.J. (1998) *Water supply and pollution control* (6th edition). Addison, Wesley and Longman Inc., Calif.

Walker, G., Gilfedder, M. and Williams, J. (1990) *Effectiveness of current salinity farming systems in the control of dryland salinity*. CSIRO publication, Camberra.

Wang, S.X. and Singh, V.P. (1995) Frequency estimation for hydrological samples with zero values. *Journal of Water Resources Planning and Management* 121:98–108.

Wanielista, M.P. (1990) *Hydrology and water quantity control*. J. Wiley & Sons, New York.

Ward, R.C. (1984) On the response to precipitation of headwater streams in humid areas. *Journal of Hydrology* 74:171–189.

Wharton, G. (1995) The channel-geometry method: guidelines and applications. *Earth Surface Processes and Landforms* 20:649–660.

Wilby, R.L. and Dettinger, M.D. (2000) Streamflow changes in the Sierra Nevada, California, simulated using a statistically downscaled general circulation model scenario of climate change. In: S.J. McLaren and D.R. Kniveton (eds) *Linking climate change to land surface change*. Kluwer Academic Publishers, Dordrecht, pp. 99–121.

Wilby, R.L., Hay, L.E., Gutowski, W.J. Jnr., Arritt, R.W., Takle, E.S., Pan, Z., Leavesley, G.H. and Clark, M.P. (2000) Hydrological responses to dynamically and statistically downscaled climate model output. *Geophysical Research Letters* **27**: 1199–1202.

Williamson, D.R., Stokes, R.A. and Ruprecht, J.K. (1987) Response of input and output of water and chloride to clearing for agriculture. *Journal of Hydrology* **94**:1–28.

Wilson, G.V., Jardine, P.M., Luxmoore, R.J. and Jones, J.R. (1990) Hydrology of a forested hillslope during storm events. *Geoderma* **46**:119–138.

Wood, W.E. (1924) Increase of salt in soil and streams following the destruction of the native vegetation. *Journal of the Royal Society of Western Australia* **10**:35–47.

Yang, J., Li, B. and Liu, S. (2000)A large weighing lysimeter for evapotranspiration and soil water–groundwater exchange studies. *Hydrological Processes* **14**:1887–1897.

Zaslavsky, D. and Sinai, G. (1981) Surface hydrology, I. Explanation of phenomena. *Journal of the Hydraulic Division, Proceedings of the American Society of Civil Engineers* **107**:1–16.

Zhao, D. and Sun, B. (1986) Air pollution and acid rain in China. *Ambio* **15**:2–5.

Zinke, P.J. (1967) Forest interception studies in the United States. In W.E. Sopper and H.W. Lull (eds) *Forest hydrology*. Pergamon, Oxford.

INDEX

RELATED TITLES AVAILABLE FROM ROUTLEDGE

Fundamentals of Geomorphology
Richard John Huggett
University of Manchester

| | | Hb: 0-415-24145-6 |
| Fundamentals of Physical Geography Series | Routledge | Pb: 0-415-24146-4 |

Fundamentals of Soils
John Gerrard
University of Birmingham

| | | Hb: 0-415-17004-4 |
| Fundamentals of Physical Geography Series | Routledge | Pb: 0-415-17005-2 |

Fundamentals of Biogeography
Richard Huggett
University of Manchester

| | | Hb: 0-415-15498-7 |
| Fundamentals of Physical Geography Series | Routledge | Pb: 0-415-15499-5 |

The Hydrology of the UK
A study of change
Edited by Mike Acreman

| | | Hb: 0-415-18760-5 |
| | Routledge | Pb: 0-415-18761-3 |

Eco-Hydrology
Edited by Andrew J. Baird and Robert L. Wilby

| | | Hb: 0-415-16272-6 |
| Physical Environment Series | Routledge | Pb: 0-415-16273-4 |

Information and ordering details

For price availability and ordering visit our website **www.tandf.co.uk**

Subject Web Address   **www.geographyarena.com**

Alternatively our books are available from all good bookshops.